How to Solve
Word Problems in Arithmetic

Phyllis L. Pullman

McGraw-Hill

New York   San Francisco   Washington, D.C.   Auckland   Bogotá
Caracas   Lisbon   London   Madrid   Mexico City   Milan
Montreal   New Delhi   San Juan   Singapore
Sydney   Tokyo   Toronto

Library of Congress Cataloging-in-Publication Data applied for.

McGraw-Hill

A Division of The **McGraw·Hill** Companies

Copyright © 2001 by The McGraw-Hill Companies, Inc. All rights reserved.
Printed in the United States of America. Except as permitted under the United
States Copyright Act of 1976, no part of this publication may be reproduced or
distributed in any form or by any means, or stored in a data base or retrieval
system, without the prior written permission of the publisher.

1 2 3 4 5 6 7 8 9 0   DOC/DOC   0 9 8 7 6 5 4 3 2 1 0

ISBN 0-07-136271-1

*The editing supervisor was Maureen B. Walker and the production supervisor was
Tina Cameron. It was set in Stone Serif by PRD Group.*

Printed and bound by R. R. Donnelley & Sons Company.

McGraw-Hill books are available at special quantity discounts to use as premiums
and sales promotions, or for use in corporate training programs. For more
information, please write to the Director of Special Sales, Professional Publishing,
McGraw-Hill, Two Penn Plaza, New York, NY 10121-2298. Or contact your
local bookstore.

 This book is printed on recycled, acid-free paper containing
a minimum of 50% recycled, de-inked fiber.

Contents

Preface

All too often, students faced with word problems seem almost ready to "give up" even before they start. In many cases, the difficulties arise NOT from a lack of knowledge, but rather from a lack of (self-)confidence that almost turns to fear. It is my hope that this book may help to take away some of the fear and self-doubt. I have tried, in many cases, to give various alternate approaches. Perhaps seeing that there is NOT JUST ONE WAY to do word problems, may not only simplify your task, but will help you to regain self-confidence in your ability.

I have also included some mathematical curiosities and tricks, so you can find some fun and success with math, also hopefully building self-confidence.

The book may be used on its own, or as a complement to the textbook.

If you want, or need, solutions for any problems from the "mixed bag," or if you are interested in more mathematical curiosities, please feel free to contact me at

Marie Curie Middle School 158Q
46-35 Oceania Street
Bayside, NY 11361

or through

McGraw-Hill
Two Penn Plaza
New York, NY 10121–2298

I hope you find the things in here helpful and enjoyable.

PHYLLIS PULLMAN

Chapter 1

Approaching Word Problems

Perhaps the first thing we must discuss is just what we mean by a "word problem." Clearly, it must be a problem that is given to us in words. But a "word problem" in mathematics also means that a whole set of skills other than those thought of traditionally as "math" (that is, "arithmetic") must be called upon.

When a student encounters word problems for the first time, it often marks the first time that, rather than being told exactly what operation(s) must be performed, the student must make that decision. This can sometimes be the most difficult part. There are, however, certain simple steps that the student can follow to help ease the task.

- **READ** the problem—<u>carefully</u>. This sounds so obvious, but all too often errors result from students simply picking out the numerical values in the problem and making their own decisions as to what to do with them.
- **ASK** yourself, "What is the problem telling me?" "What am I looking for?"
- **ANALYZE.**
- **THINK** about how you might be able to accomplish this, remembering that often you have a variety of methods you can use. (More is said about these choices a little later.) And, if you cannot figure out a plan of action, try to set up a "parallel" situation, perhaps using numerical values, or even just a situation more familiar to you.
- **PLAN.**

- **DRAW** a diagram, if appropriate, and **COPY** all information carefully and correctly.
- **WORK** it out, again making sure to work neatly and accurately.

When you have your solution, always make sure to ask yourself a very simple, but very important, five words: **DOES MY ANSWER MAKE SENSE?** (For example, a down payment should not be more than the total cost of an item; sales tax should not be more than the cost of the item; if we are looking for a number of people involved in doing something, we cannot [or, certainly, should not] find anything but a whole number; etc.)

Finally, before you "give your answer," **CHECK** in the conditions of the problem. (If you have used algebra to solve your problem, DO NOT CHECK IN THE EQUATION. Go back to the <u>words</u> of the original problem. After all, if <u>you</u> created the equation, it may not have been a correct one, even though you may have solved it correctly.) Mathematics is a subject you learn by <u>doing</u>, not by reading. The more problems you do, the easier they will become.

As mentioned earlier, there are a variety of strategies you can use. So often, my students ask, "Which one is the 'right' one?" or "Which one should I use?" Very simply, there is no "right" one, and you may even find that if called upon to solve the same problem on more than one occasion, you may even use different strategies each time!

<u>Some</u> of the alternative approaches (it really is impossible to list <u>all</u>!) are:

- Patterns
- Working backward
- Trial and error (and knowing what to rule out, as well as what to rule in)
- Shortcuts
- Considering parallel situations
- Algebraic
- <u>Explaining</u> a solution

2

- Estimation and approximation
- Using technology

Now, what does each of these mean?

Patterns. Very often, if a problem seems to be just a great deal of "busy work," often involving repeated computations, start slowly and don't attempt to "do the whole thing at once." Rather, do the computations one at a time, and examine each result. You may suddenly find that you will be able to predict an answer that might seem to be "way down the line" by noting the pattern and seeing where the required answer fits into that pattern.

For example, even a problem that seems very cumbersome, such as finding the units digit of $(123,456,789,123)^{98}$ becomes very simple when the "right approach" is used. First of all, a realization that the only one of the 12 digits that can affect the units digit of the result immediately "reduces" the problem to 3^{98}. If you then work one power of 3 at a time, you see something very interesting happening.

$$3^1 = \underline{3}$$
$$3^2 = \underline{9}$$
$$3^3 = 2\underline{7}$$
$$3^4 = 8\underline{1}$$
$$3^5 = 24\underline{3}$$

Aside from the fact that it isn't really even necessary to find the actual values of these powers (the units digit is sufficient), we see that a "pattern" has emerged. The units digits will be 3, 9, 7, 1, and will then repeat again and again ($24\underline{3} \times 3 = 72\underline{9}$; $72\underline{9} \times 3 = 218\underline{7}$; etc.). Now, all we have to do is to find where in this cycle the 98th power would appear. You can then divide by 4, and find that there is a remainder of 2. Therefore, the 98th power would be in the "second slot," and would therefore have a units digit of **9.**

Working Backward. Sometimes it is possible, by knowing the "end result," to "undo" what has been done.

For example, Mrs. Jones decided to give donations to several of her favorite charities. She allotted a certain amount of money. She decided to give half to Meals on Wheels, one-fourth of what remained to Toys for Tots, $100 to Penny Harvest, and the $50 that remained to her college alumni fund.

How much money had she allotted for charities this year? Start with the final donation, the amount that "remained":	$50
Add the previous donation of $100:	$150
This represented <u>three</u>-fourths of what remained after the First donation. Divide by 3 to find the "remaining fourth" ($50) And add it to the $150:	$200
This represents the "half that was left" after the initial donation. Therefore, the original amount had to be double this—or—	**$400**

Trial and Error. Make a guess. (Try to make an "educated" guess—that is, a realistic one—to save yourself some work, but if you cannot, make any choice.) Then proceed from this guess, until you find the numbers which will satisfy all conditions in the problem.

For example, I am thinking of two numbers. One number is 25 more than the other. The sum of the numbers is 75. Find the numbers.

Try 30. 25 more than this is 55. $30 + 55 = 85$. But we want a sum of 75. Our sum is 10 more than we want.

You might then think to "split" the 10, and try 25. 25 more than this is 50. $25 + 50 = 75$. This satisfies the second condition, and, therefore, our numbers are **25** and **50.**

(If you did not realize to "split" the 10, but simply tried 10 less than your original choice, you would try 20. $20 + 25 = 45$. $20 + 45 = 65$. This is 10 less than we want. Since 30 gave us 10 more than we wanted, and 20 gave us 10 less than we wanted, the next logical choice would be midway between 30 and 20, or 25, leading us to the same result as above.)

Shortcuts could include tests for divisibility, using "common sense" and logic, as well as others to be discussed more fully, when appropriate. For example, if numbers are divisible by 3, the sum of the digits must be divisible by 3, and if numbers are divisible by 9, the sum of the digits must be divisible by 9. So, if you were to want to know what the value of A would have to be if **834682A6** is to be divisible by 18, you could actually substitute each of the values from 0 through 9 until you can divide by 18 without getting a remainder (other than 0), BUT, it is so much easier if you realize that a number divisible by 18 must be divisible by 2 (it must be even; this is) and by 9. Since a number divisible by 9 must be such that the sum of the digits is a multiple of 9, first find the sum of the digits. If you add $8 + 3 + 4 + 6 + 8 + 2 + 6 = \underline{37}$. The next multiple of 9 is $\underline{45}$. What would you have to add to 37 to get 45? **8.** Therefore, the value of A must be **8.**

Considering Parallel Situations.   This can be a particularly effective and useful method when you are faced with a situation where you face a problem presented in the abstract, and you are not sure what operation must be performed. For example, if you are asked to express the number of inches in X feet, and you know that you have to introduce 12, BUT you are not sure whether you should multiply or divide. If you simply ask yourself how many inches there are in 2 feet, you would probably almost automatically answer 24. What did you do? You multiplied. Therefore, the number of inches in X feet must be **12X.**

The other alternatives will be more thoroughly discussed as appropriate.

Chapter 2

Reviewing the Basics

No shortcuts or "tricks" to make solution of verbal problems simple will be of any use if you cannot do the basics. Calculators can help, of course, but we do not always have one available. In addition, too much reliance on a calculator can "prove hazardous to our [mathematical] health"; a four-function calculator, for example, does not even follow the order of operations! It is particularly unfortunate to see a student change a correct answer to an incorrect one because "the calculator said so."

In these pages, we will review operations involving integers and the order of operations. There are, of course, "rules" for performing the various operations with integers, or "signed numbers." For students who have trouble remembering the "rules," there are some alternative strategies that can be used. Let's examine the operations one at a time.

Addition

Addition of two numbers with the same sign seldom causes problems. Addition of positive integers never seems to create problems; they are what we have always worked with as numbers of arithmetic. Addition of two negative numbers sometimes creates problems (particularly when students confuse the rules for another operation). Setting the problems in real life situations always seems to make such problems much, much simpler. Create a real life situation involving something you know about: owing money and owing more money; a football player who loses yards on one play and then loses additional yards on the next play; walking down some stairs and

7

then down some more stairs; giving away so many of an item and then giving away more of that item; etc.

Using the same strategy when adding numbers with different signs generally makes what could be a difficult problem very simple. Most students, in fact, were able to deal with adding numbers with different signs long before they ever heard of "integers," "signed numbers," or even "positive" or "negative." You may not remember the "rules"; you may not be able to remember how to handle such a problem, BUT, think about money your friend owes you and money you owe your friend, and invariably you'll know immediately who has to pay, who gets to collect, and just how much! If that doesn't work, try thinking of walking up some stairs and then down. You'll be able to figure out whether you are above or below where you started. Hopefully, these will help you see the "why" behind the rules. When you know the "why," remembering rules becomes understanding, NOT memorization.

Adding numbers with the same sign: Keep the sign and find the sum of the absolute values.

Adding numbers with different signs: Keep the sign of the addend with the greater absolute value and find the difference of the absolute values.

(If it helps, try thinking **same**...**sum** and **different**... **difference**.)

Subtraction

The easiest way, of course, to handle subtraction is to remember that **subtraction of a signed number is equivalent to the addition of its opposite.** If you need help understanding why this "works," you need only consider some problems using a number line and the "what do I have to add to get what I want, or back to where I started" concept of subtraction. Almost everyone, at one time or another, has received change from someone who does it without ever calculating exactly how much to give the person. Rather, the person making change starts with the cost of your item, and then starts adding to it until he or she has reached the amount you tendered in the first place. (For example, you buy an item for $1.47 and pay for it with a $5 bill. The cashier then hands you 3 cents and says, "$1.50," hands you 2 quarters and says, $2, and then proceeds to hand you three more dollar bills, and says, "$3, $4, $5".)

If you consider some subtraction problems using this concept and a number line, and then look back at some carefully chosen addition problems where you have added the opposite of the number subtracted in the subtraction problems, you will see the validity of the rule.

Multiplication

I always like using real life situations to help develop the rules for multiplication too. Just consider depositing money, gaining weight, or earning money as positive situations, and withdrawing money, losing weight, or paying wages as negative situations. Then think of time in the future as positive, and time in the past as negative, and make up some simple problems that will help you easily remember the "rules."

For example, if you **deposit** $20 a week in your savings account, 5 weeks **from now** you will have $100 **more** in the account. Or, if you **gain** 2 pounds a week, 4 weeks **from now** you will weigh 8 pounds **more** than you do now. Similarly, if you **earn** $10 a week, 7 weeks **from now** you will have $70 **more** than you have now (**positive × positive = positive**). If you **deposit** $20 a week in your savings account, 5 weeks **ago** you had $100 **less** in the account, since the money had not yet been deposited. If you have been **gaining** 2 pounds a week, 4 weeks **ago** you weighed 8 pounds **less** than you do now, since you had not yet gained the weight. If you are **earning** $10 a week, 7 weeks **ago** you would have not yet earned the money and would therefore have $70 **less** than you do now (**positive × negative = negative**).

If you **withdraw** $20 a week from your savings account, 5 week **from now** you will have $100 **less** in your account, since you will have withdrawn that amount of money. If, like most people, your diet is one to lose weight, you **lose** 2 pounds a week, 4 weeks **from now** you will have lost the weight and will weigh 8 pounds **less** than you do right now. If you have hired someone to work for you and you must pay that person, it will cost you whatever money you have to pay out. So, if you

pay someone $10 a week, 7 weeks **from now** you will have $70 **less** than you do right now, because you will have paid out that amount of money **(negative × positive = negative).**

And finally (and this always seems to be the one most difficult to accept) if you **withdraw** $20 a week from your bank account, 5 weeks **ago** you had **more** money in your account, since you had <u>not yet withdrawn the money</u>. If you're on that diet, and you are **losing** 2 pounds a week, 4 weeks **ago** you would not yet have lost the weight and therefore would weigh 8 pounds **more** than you do right now. And, finally, if you are **paying** someone $10 a week, 7 weeks **ago** you would not yet have paid the person, and would therefore have $70 **more** than you do right now **(negative × negative = positive).**

Since determining the sign of your product is the only really new concept to remember here, remembering these results should enable you to do any multiplication problem you might get. Or there is always the story of the good guys and the bad guys. (It's a silly story, but my students like it, and maybe they find it easy to remember BECAUSE it is silly.) Here all you have to remember is the "good" is positive and "bad" is negative. So,

If something **good** happens to the **good** guy...that's **good** (positive × positive = positive).
If something **good** happens to the **bad** guy...that's **bad** (positive × negative = negative).
If something **bad** happens to the **good** guy...that's **bad** (negative × positive = negative).
And, if something **bad** happens to the **bad** guy...that's **good!** (negative × negative = positive).

Division

The "rules" for division are the same, but if you need to understand "why" to help you remember, all you have to do is to remember that, for example, $8/2 = 4$, because 4 is the number which when multiplied by 2 gives you 8. If you think of division in terms of the related multiplication problem, when you divide a positive by a negative, the quotient must be such that when you multiply it by the negative divisor, the result is positive. Since the signs must be the same in order to get a positive product, the quotient must be negative, too **(positive ÷ negative = negative).**

Similarly, when you divide a negative by a positive, the quotient must be such that when you multiply it by a positive, the result is negative. Since the signs must be different in order to get a negative product, the quotient must be negative **(negative ÷ positive = negative).**

Finally, when you divide a negative by a negative, the quotient must be such that when you multiply it by a negative, the result is negative. Again, since the signs must be different in order to get a negative product, the quotient must be positive **(negative ÷ negative = positive).** (I've left out the "obvious," since divisions of two positives is simply division of two "numbers of arithmetic" and must result in another "number of arithmetic."

Order of Operations

I'm not really sure who it was who first suggested that "**P**lease **E**xcuse **M**y **D**ear **A**unt **S**ally" would be a helpful way to remember the correct order of operations, but it certainly has proved to be a very valuable aid to many students. There is one problem with using "**PEMDAS.**" Since no two things can occupy the same space simultaneously, there is no way to, indicate that multiplication and division are "on the same level" (that is, that neither one takes priority over the other). A similar problem exists for addition and subtraction. And <u>remember</u>, **a four-function calculator neither knows nor is it able to follow the "order of operations."**

Chapter 3

Problems Involving Perimeter and Area

It has been said that "one picture is worth 10,000 words." In the case of problems involving perimeter and area, that "picture" (the diagram) can in fact be worth even more than that.

First, of course, we must understand the meaning of, and the difference between, perimeter and area. **Perimeter** literally means "distance around," and therefore it is the sum of the lengths of the sides of the figure. It is **one**-dimensional. **Area** is the measure of the surface of a plane figure, or the number of square units it contains. It is **two**-dimensional.

Before we solve word problems involving perimeter and area, however, it is important to review the formulas for finding the perimeters and areas of the various geometric shapes. When it comes to area, I think, too, that it is important to understand the derivation of these formulas (where they come from, and why). The problem many people face with area formulas is an inability to "memorize" them. If you understand where the formula came from, however, you will not have to rely on your memory.

Finding the area of a rectangle can be very simple if you draw the figure on graph paper and then "count" the number of square units contained in it. From this counting procedure, it is very easy to see that the number of square units is simply the number of squares in one row multiplied by the number of rows. Since the number of units in one row is simply the units in the length (or base) of the rectangle, and the number of rows in the units in the width (or height) of the rectangle, we can easily see that to find the area of a rectangle, we must multiply the length and width (or base and height). The formula, then, is $\mathbf{A = lw}$ or $\mathbf{A = bh}$. Using what we know about the rectangle, we can easily find the area of a parallelogram. A parallelogram also has a base and a height (as indicated in the diagram below).

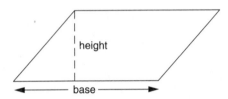

If we cut along the height of the parallelogram and move the triangle to the "other side," notice that we now have another rectangle. This rectangle has the same base and same height as the original parallelogram, and also **has the same area.**

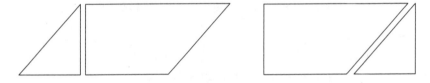

Since the dimensions are the same, and the areas are the same, we must be able to find the area the same way. Therefore, for a parallelogram, $\mathbf{A = bh}$. If we now take our parallelogram and draw a diagonal, notice that two triangles are formed. How do the two triangles compare to one another? Are they the same shape? Are they the same size? (If you are having trouble visualizing the relationship, you can always cut out a parallelogram, draw a diagonal, and then cut along the diagonal.) What you should have found is a situation that looks like this:

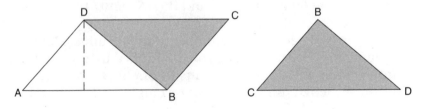

Since the two triangles have the same area, each must have an area equal to ½ that of the parallelogram. Therefore, the formula must be **A = ½ bh.**

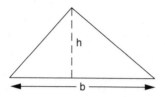

Similarly, we can derive a formula for finding the area of a trapezoid by again relying on what we know about the parallelogram. Draw a trapezoid, and cut it out. Now cut out a second trapezoid, identical to the first. (If you don't have scissors handy, try to draw and visualize.)

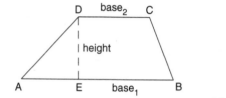
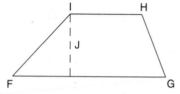

Now, "flip" the second one, horizontally and vertically, and place it next to the first in such a way as to make the legs coincide, forming a parallelogram.

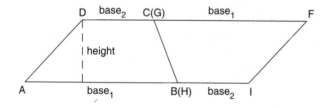

Now let us consider the "base" and "height" of this new parallelogram.

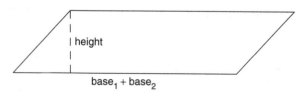

The "base" is equal to the sum of the lengths of the bases of the trapezoids, or $b_1 + b_2$. The "height" of the parallelogram is exactly the same as the height of the original trapezoid. Since the area of a parallelogram is $A = $ base $\times$ height, the area of this "parallelogram" is $(b_1 + b_2) \times h$. However, remember that the parallelogram was formed using <u>two</u> of our trapezoids; therefore, the area of the trapezoid must be <u>½ that of the parallelogram</u>. That means that the formula for finding the area of a trapezoid must be $A = ½(b_1 + b_2)h$ or, if you prefer, $A = ½ h(b_1 + b_2)$.

LET'S SUMMARIZE:

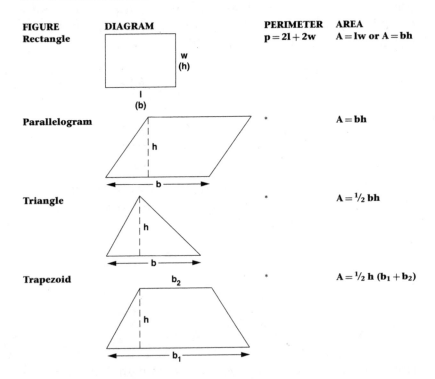

FIGURE	DIAGRAM	PERIMETER	AREA
Rectangle		$p = 2l + 2w$	$A = lw$ or $A = bh$
Parallelogram		*	$A = bh$
Triangle		*	$A = ½ bh$
Trapezoid		*	$A = ½ h(b_1 + b_2)$

Let's try using some of these formulas in some problems concerning these polygons, and then we will look at the circle.

*Determine the perimeter by finding the sum of the lengths of the sides.

MODEL PROBLEM I

Johnny would like to plant some bulbs in his garden. Each bulb needs 2 square feet. If his garden looks like the diagram below, how many bulbs can he plant?

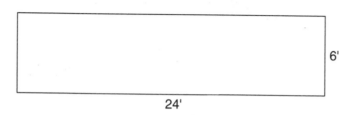

24'

First we have to decide whether we are looking for the perimeter or the area. Since each bulb needs 2 square feet, we must find the **number of square feet in his garden,** or in other words, the **area.** This time we do not have to worry about a diagram, since one is given to us. Therefore, what we must determine first is the **shape** of the garden, and what **formula** we must use.

The garden is in the shape of a rectangle. Therefore, **A = bh.** Once you have determined the formula to use, **substitute** the information you have in that formula.

$$A = bh$$
$$A = 24 \times 6$$
$$A = 144 \text{ square feet or } 144 \text{ ft}^2$$

For every **2** of those square feet, Johnny can plant a bulb. Therefore, we have to **divide** our area by 2, and we find that Johnny will be able to plant 72 bulbs.

MODEL PROBLEM 2

Suppose that Johnny wants to put a fence around this garden. How many feet of fencing will he need?

There are two clues here that we are looking for the perimeter. This fence will go **around** the garden, and we are looking for the number of **feet** (a linear measure). Ideally, you will remember the formula for finding the perimeter of a rectangle, BUT, if you don't, remember that the perimeter is simply the **sum of the lengths of the sides.** (And one of the reasons a diagram is so important here is that hopefully it will prevent you from making what is probably the most common error in perimeter problems—"forgetting" one or more of the sides of the figure!)

$$\mathbf{p = 2l + 2w}$$
$$p = 2(24) + 2(6)$$
$$p = 48 + 12$$
$$p = 60 \textbf{ feet}$$

Suppose this fencing will cost Johnny $7.50 per foot. How much will he have to pay for the fencing?

Each foot will cost $7.50.

Johnny has **60** feet.

Therefore, Johnny will have to pay $7.50 × 60, or **$450.**

MODEL PROBLEM 3

Suppose that Johnny's fence is a wooden one 5½ feet high. Johnny decides that he does not like the natural color of the wood, and wants to paint the fence. He goes to the paint store and is told that 1 gallon of paint will cover 22 square feet of wood. How much paint will he have to buy?

Let's draw the diagram and look for clues as to what we have to find and how we can accomplish that.

The fence:

$5\frac{1}{2}'$

60'

To paint the fence means we have to cover the **surface** of the fence. This should help us to recognize that we have to find the area. In addition, we are told that the paint will cover a certain number of **square** feet. **Square** units measure **area.**

$$A = \mathbf{bh}$$
$$A = 60 \times 5\frac{1}{2}$$
$$A = \mathbf{330 \text{ square feet}}$$

(In case you need some review to multiply $60 \times 5\frac{1}{2}$, there are several ways you can do the problem.

$$60 \times 5\frac{1}{2} = 5 \times 60 + \frac{1}{2} \times 60 = 300 + 30 = 330$$

OR

$$60 \times 5\frac{1}{2} = 60 \times \frac{11}{2} = 30 \times 11 = 330$$
$$\left[\text{or, if you don't want to simplify: } \frac{660}{2} = 330 \right]$$

OR if you are like so many people and consider yourself "allergic" to fractions, $5\frac{1}{2} = 5.5$ and therefore $60 \times 5\frac{1}{2} = 60 \times 5.5 = 330$).

Since a gallon of paint will cover 22 square feet, we need to know how many sections of 22 square feet we have, or, in other words, we have to divide 330 by 22. $330/22 = 15$. Johnny will need **15** gallons of paint.

(Remember, in real life problems, sometimes the mathematical answer is not the final one—or the answer to the question. If, for example, when we divided, we had a quotient of something other than a whole number, **we would have had to answer the next whole number,** since you cannot buy part of a gallon of paint.)

MODEL PROBLEM 4

A parking lot has spaces in the shape of parallelograms. Each parallelogram has a base of 19 feet and an altitude of 9 feet. What is the area of one space? If the parking lot has 40 such spaces, what is the total parking area available?

In this problem, two possible questions have already been answered. The figure involved is a **parallelogram,** and we are looking for the **area.**

Write the formula: $A = bh$
Substitute in the formula: $A = 19 \times 9$
Evaluate: $A = $ **171 square feet**

This is the area of one parking space. To find the total area devoted to parking, just multiply the area of one space by the total number of spaces. (Notice that the problem did NOT ask for the total area of the lot. That would involve the lanes for driving, or any other section not used for the actual parking.)

Total area $= 171 \times 40 = $ **6840 square feet**

Sometimes area problems require **strategy.** Real life does not always come in specific, known shapes.

MODEL PROBLEM 5

Find the area of the following figure:

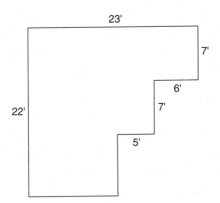

There, of course, is no formula for such a shape. BUT we can (easily) break it down into known shapes, for which we have formulas. (There is also not necessarily ONE correct way to do the problem; it all depends on how you see the figure at the moment you are doing the problem.)

20

Here are **three** ways this problem can be done:

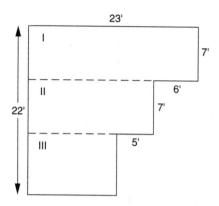

Draw three horizontal line segments, dividing the figure into three rectangles. Find the area of each rectangle.

Rectangle I:

$$A = bh$$
$$A = 23 \times 7$$
$$A = \textbf{161 square feet}$$

Rectangle II: We know the height is 7 feet, but we must first **find** the base. Since the base of rectangle I is 23 feet and rectangle II is 6 feet shorter than rectangle I, the base of rectangle II is 23 − 6, or 17 feet. Therefore,

$$A = bh$$
$$A = 17 \times 7$$
$$A = \textbf{119 square feet}$$

Rectangle III: We aren't told either dimension of this rectangle, but we can figure them out easily. We can determine the base using the same strategy we used for rectangle II: 17 − 5 = 12 feet. To find the height of rectangle III, we can use the fact that the entire figure is 22 feet high, and rectangles I and II account for 7 + 7, or 14 feet. Rectangle III is therefore 22 − 14, or 8 feet high.

Therefore,

$$A = bh$$
$$A = 12 \times 8$$
$$A = \textbf{96 square feet}$$

The **TOTAL AREA** of the figure is $161 + 119 + 96$, or **376 square feet.** If we had preferred, we could have drawn vertical line segments initially. The figure would then have looked like this.

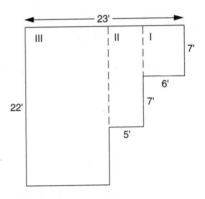

This time we'll number the rectangles from right to left, because of the information we have (and we'll leave the "figuring" for last).

Rectangle I:

$$A = bh$$
$$A = 6 \times 7$$
$$A = \textbf{42 square feet}$$

Rectangle II: We know that the base of this rectangle is 5 feet. The height is simply the combined heights of the first and second sections, or $7 + 7$, or 14 feet. Therefore,

$$A = bh$$
$$A = 5 \times 14$$
$$A = \textbf{70 square feet}$$

Rectangle III: This time we know that the height is 22 feet, and we must find the base. Since the total figure is 23 feet across, and sections of 5 feet and 6 feet have been accounted for, this rectangle has a base of $23 - (5 + 6)$ or $23 - 11$, or 12 feet. The area of this rectangle then is

$$A = bh$$
$$A = 12 \times 22$$
$$A = \textbf{264 square feet}$$

The **TOTAL AREA** is $42 + 70 + 264$, or **376 square feet** (as expected, exactly the same as it was when we drew the horizontal line segments instead). You might even prefer to make believe you have a larger figure and "cut away" the parts you don't want. In that case, the diagram would look like this.

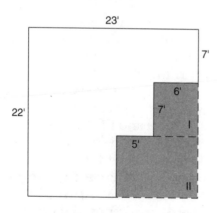

The area of the large rectangle is

$$A = bh$$
$$A = 23 \times 22$$
$$A = \textbf{506 square feet}$$

However, we do not want the shaded area. Therefore, we will have to "cut it away," or, in mathematical terms, SUBTRACT that area. Since in this case the figure whose area we must subtract is NOT a figure for which there is a formula, in this case we will have to "break it up," as we did in the first two methods.

Rectangle I:

$$A = bh$$
$$A = 6 \times 7$$
$$A = \textbf{42 square feet}$$

Rectangle II: The base of this rectangle is $5 + 6$, or 11, feet. We can find the height the same way we already did in the first method. Therefore,

$$A = bh$$
$$A = 11 \times 8$$
$$A = \textbf{88 square feet}$$

That means that we must cut away areas of $42 + 88$ or **130 square feet.** Therefore, the desired area is $506 - 130$ or **376 square feet.** It's time now for some practice problems, but before you start, a warning. Be very careful, when you are doing **area** problems, that you do not try to use any **extraneous** information they may give you. Finding the area generally requires knowing the **base** and the **height.** If you are given the lengths of any other sides of the figure, IGNORE THEM (unless you need to find the perimeter too!)
Remember to draw a diagram (if you are not given one); then determine what kind of figure you have (if you are not told), and then decide whether you are looking for the <u>perimeter</u> or the <u>area</u> and select the appropriate <u>formula</u>.

Practice Problems Involving Perimeter and Area

1. Mrs. Simons wants to put down ceramic tiles in her entry hall. The hallway is in the shape of an L, where each section of the L is a rectangle. One rectangle is 11 feet by 6 feet, and the other is 4 feet by 15 feet. If each ceramic tile is 1 square foot, how many of these tiles will she need?

24

2. Mr. Robbins is going to paint the back of his barn. It is in the shape of a rectangle, topped by a triangle. The back of the barn is 35 feet from side to side. If the height of the "rectangle" is 11 feet and the height of the triangle is 4 feet, find the total area he has to paint. If each gallon of paint covers 4 square feet, how many gallons does he have to buy? If each gallon costs $12.75, how much will the paint cost him?

3. The local fish hatchery is going to rope off an area where all the newborn fish will be kept. If this area is in the shape of the trapezoid shown on the right, how many square feet will the fish have for swimming?

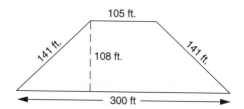

105 ft.

141 ft.   108 ft.   141 ft.

← 300 ft →

4. Mr. Jeffries is afraid that someone will slip on his pool deck. In order to avoid that problem, he has decided to put down a "skidproof carpet" around the pool. It will be cut to order, and he only has to pay for what he uses. The pool and deck are illustrated to the right. If this special "carpet" costs $25 a square foot, how much will he have to pay?

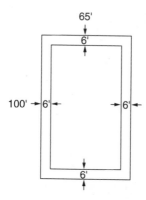

65'
6'
100' → 6' ←   → 6' ←
6'

5. Mr. Rogers knows that it should take 1 hour to plow and plant every 1000 square yards of his field. If he has a rectangular field which is 500 yards by 225 yards, how long should the job take?

6. Mrs. Samson has decided to refinish a bookcase. She has to refinish the back and the two sides. If the bookcase is 5 feet across, 3 feet deep, and 5 feet high, what area will she have to cover?

7. Delilah got a part-time job mowing lawns. When she couldn't decide how much to charge, her younger brother came up with an idea. He suggested that her customers pay according to the area she has to mow. She will get $0.05 for every square foot she mows. If the people for whom she is working have a backyard that is in the shape of a rectangle 60 feet by 45 feet, how much will she earn?

8. The people for whom Delilah is working have decided to put a picket fence around the yard. The picket fence costs $8.50 per foot. How much will it cost them?

9. How many square inches are there in a square foot?

10. How many square feet are there in a square yard?

11. How many square inches are there in a square yard?

Solutions to the Practice Area and Perimeter and Problems

1. First of all, let's draw a diagram. Since each of the two sections is a rectangle, the formula will be the same: **A = bh.**

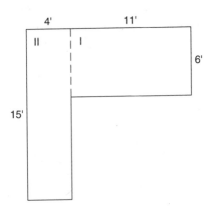

For rectangle I, $b = 11$ feet and $h = 6$ feet.

Therefore, $A = bh$

$A = 11 \times 6$

$A = $ **66 square feet**

Surface area of a three-dimensional sphere = 4 × π × radius²:

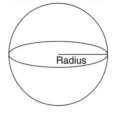

Measuring Volume

Volume of a rectangular solid = length × width
× height or area of the base × height:

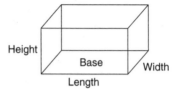

Volume of a cube = side³:

Volume of a sphere = ⁴⁄₃ × radius³

Volume of a circular cylinder = π × radius² × height:

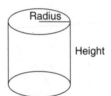

Volume of a circular cone = ¹⁄₃ × π × radius² × height:

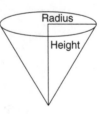

Money-Related Formulas

Total dollar amount of item = quantity of item × dollar amount of one item
Total cost = price + tax rate × price
Discounted price = original price − discount × original price
Interest earned = principal × rate of interest × time invested
Current amount = principal + interest earned

Chapter 2

Using Algebraic Equations to Solve Problems

In the previous chapter, you studied problems that were solved by making basic arithmetic computations. In this chapter, you will learn how to apply algebra to solve problems. In most cases, this will require *modeling* the situation with an equation in which what you wish to find will be expressed by a letter representing an unknown quantity. We call this the *variable* of the equation. The equations will come from applying basic formulas and well-known concepts. (You can review basic tools of algebra by visiting the Appendix in the back of the book.)

It is important to carefully identify the *unknown variable* in the problem and determine the relationship between it and the other information in the problem. It is also important that your answer really does solve the problem! There are many places to make mistakes in these problems, such as creating the wrong equation, solving the equation incorrectly, or making an arithmetic mistake.

Problems Involving Numbers

The simplest of all word problems are those that involve finding numbers. Most of these problems will tell you how two or more numbers are related. The key will be to identify all the relationships between the numbers that are mentioned in the problem.

Example 1

The larger of two numbers is equal to 14 more than 3 times the smaller number. Find the smaller number if their difference is 38.

Solution 1

The first step of any word problem is to find a way to use a variable to represent the unknowns in the problem. Since the problem specifically tells us how the larger and smaller numbers are related, we can use one variable. It is also a good idea to select a letter for the variable that reminds us of what we're looking for. The letter n is always a good choice for a number.

The first sentence of this problem tells us that if we know the smaller number, we can compute the larger one. It makes sense, therefore, to use the variable to represent the smaller one. We should also start our work with a statement of how we are assigning variables. This is often called the "Let" statement.

$$\text{Let the smaller number} = n$$

We are now ready to start translating the first sentence into an algebraic equation. A good idea is to first rewrite the problem leaving most of the words and using mathematical symbols for adding, subtracting, multiplying, dividing, and equality when it is easily recognized. This would give us

$$\text{Larger number} = 14 \text{ more than } 3 \times \text{smaller number}$$

Now we have to take a look at the phrase on the right. We must realize that to have *more than* something means to *add* to that something.

$$\text{Larger number} = 3 \times \text{smaller number} + 14$$

We can now substitute our variable expressions for the smaller and larger parts. Remember to drop the multiplication symbol when multiplying by a variable.

$$\text{Larger number} = 3n + 14$$

Both numbers in the problem are now represented by variables, and we are ready to form the equation that we will use to solve the problem. This will come from the other relationship given in the problem, "their difference is 38." Again we should first translate the operation symbols and write the remaining words. We should realize that finding a difference means to subtract, and if the difference is positive, we are taking the smaller from the larger. We get

$$\text{Larger number} - \text{smaller number} = 38$$

Substituting our variables gives us

$$3n + 14 - n = 38$$

Simplifying and solving for n, we get

$$2n + 14 = 38$$
$$2n = 24$$
$$n = 12$$

Therefore, the smaller number is 12, and the larger number is $3 \times 12 + 14 = 36 + 14 = 50$. To check, we need only verify that their difference is 38. Clearly, $50 - 12 = 38.\checkmark$

Sometimes one of the relationships regarding the numbers is given indirectly; that is, it might be hidden in a key phrase.

Example 2

Separate 70 into two parts so that 5 times the smaller part is equal to 14 more than twice the larger part.

Solution 2

The problem is really asking for two different numbers whose sum is 70. We can assign a variable, n, to the smaller part. The larger part must be what is left when n is taken away from 70, $70 - n$. The "Let" statement is

$$\text{Let } n = \text{smaller part}$$
$$70 - n = \text{larger part}$$

Translating only the operations gives us

$5 \times$ smaller part $= 14$ more than twice the larger part

Now we have to take a look at the phrases on the right: specifically, "more than" and "twice." We must realize that to have *more than* something means to *add* to that something.

$5 \times$ smaller part $=$ twice the larger part $+ 14$

Now we use the fact that to have twice something means to multiply that something by 2.

$5 \times$ smaller part $= 2 \times$ larger part $+ 14$

Now we are ready to substitute our variable expressions for the smaller and larger parts. Remember to drop the multiplication symbol and to *use parentheses* when we need to multiply by an expression with more than one term.

$$5n = 2(70 - n) + 14$$

Solving in the usual way, we get

$$5n = 140 - 2n + 14$$

Add $2n$ to both sides to get

$$7n = 154$$
$$n = 22$$

The smaller part is 22, and the larger part is $70 - 22 = 48$.
The check is easy. Simply follow the numbers and words of the problem, 5 times the smaller is $5 \times 22 = 110$; 14 more than twice the larger is $2 \times 48 + 14 = 96 + 14 = 110.\checkmark$

Alternative Solution 2

Another way to solve the problem is to use two different variables to represent both the parts of 70. When we use two

32

variables, we need two equations. (See the Appendix for a review of how to work with equations with two variables.)

$$\text{Let } x = \text{the smaller number}$$

$$\text{Let } y = \text{the larger number}$$

The first equation comes from the fact that the sum of the numbers is 70.

$$x + y = 70$$

The second equation comes from the way these parts are related. The translation should still be performed as outlined above, but now we use the variables x and y to substitute for the smaller part and for the larger part.

$$5x = 2y + 14$$

We work with this pair by solving the first equation for y

$$y = 70 - x$$

and substituting in the second equation

$$5x = 2(70 - x) + 14$$

Note that we have the exact same equation to solve (even though the variable here is x instead of n) and we should arrive at the solution, $x = 22$ and $y = 48$.

The next example shows you another way to express one of the relationships indirectly.

Example 3

In this example, 5 less than 6 times the largest of three consecutive positive integers is equal to the square of the smaller minus 2 times the middle integer. Find all three integers.

Solution 3

It is clear that we are dealing with three different numbers. The key phrase in this problem is *consecutive positive integers*. To understand this, let's look at each word. The numbers have to be integers. This is the set of negative and positive whole numbers and zero $\{\ldots, -3, -2, -1, 0, 1, 2, 3, \ldots\}$; that is, we won't accept any answers that are fractions, decimals, or roots. Since the numbers are positive, we are looking only for numbers from the set $\{1, 2, 3 \ldots\}$. The most important word in the phrase is *consecutive*. This means that the three numbers are all one apart. We can then represent our numbers with the same variable, *n*.

Let n = the smallest integer

$n + 1$ = the middle integer

$n + 2$ = the largest integer

Now we can use our translation scheme. First only the symbols

5 less than 6 × largest = square of the smallest

− 2 times the middle

Less than something means to take away from that something

6 × largest − 5 = square of the smallest − 2 × middle

Square of means to raise the number to the second power

6 × largest − 5 = smallest^2 − 2 × middle

Now we can substitute our variables and remove the multiplication symbols.

$$6(n + 2) - 5 = n^2 - 2(n + 1)$$

This is a *quadratic equation* since we have an equation with a power of 2. (See the Appendix for a review of solving quadratic equations.) It must be simplified and have 0 on one side in order to solve it by factoring.

$$6n + 12 - 5 = n^2 - 2n - 2$$

To both sides of the equation, we subtract n^2, add $2n$, and add 2, to arrive at

$$-n^2 + 8n + 9 = 0$$

We can now multiply every term in the equation by -1 to get

$$n^2 - 8n - 9 = 0$$

We can now factor and solve for n.

$$(n - 9)(n + 1) = 0$$

Therefore

$$n = 9 \qquad \text{or} \qquad n = -1$$

Remember, we are looking for *positive* integers, so we can *reject* -1 as an answer. Since n was the smallest, the three integers are 9, 10, and 11.

To check, we simply use the numbers and follow the words of the problem: 5 less than 6 times the largest is $6 \times 11 - 5 = 66 - 5 = 61$.

The square of the smallest minus 2 times the middle is $9^2 - 2 \times 10 = 81 - 20 = 61.\checkmark$ (Not all consecutive integer problems will give rise to quadratic equations as you will see in the supplementary exercises.)

Problems Involving Age

Problems that involve the ages of two or more people are very similar to number problems. (After all, an age is a number.) Since we need to have two relationships, these problems may cause you to consider ages at the present time and at another time in the past or future.

Example 4

Sandy is 4 times as old as Robby is now. In 10 years, Sandy will be 6 years older than Robby. How old are they now?

Solution 4

Since we are told how to find Sandy's age if we know Robby's age, we should use a variable for Robby's age and determine the symbolic representation of Sandy's age with the same variable. Following the procedure outlined in the previous problems, we have

$$\text{Let Robby's age} = x$$
$$\text{Sandy's age} = 4 \times \text{Robby's age}$$
$$\text{Sandy's age} = 4x$$

"In 10 years" means that we will add 10 to each of their ages. We ought to state what their ages will be.

$$\text{In 10 years, Robby's age will be } x + 10$$
$$\text{and Sandy's age will be } 4x + 10$$

We should now translate the second sentence using these future ages and substitute the future variables

$$\text{Sandy's age} = \text{Robby's age} + 6$$
$$4x + 10 = x + 10 + 6$$

Simplifying and solving for x, we get

$$4x + 10 = x + 16$$
$$3x = 6$$
$$x = 2$$

Therefore, Robby's current age is 2 and Sandy's current age is 8.

To check, we should simply follow the words of the problem. Sandy's current age is 4 times Robby's age or $8 = 4 \times 2$.✓

In 10 years, Sandy will be 18 and Robby will be 12. Clearly, Sandy will be 6 years older than Robby.✓

Example 5

Jessica is 27 years old and Melissa is 21 years old. How many years ago was Jessica twice as old as Melissa?

In this problem, the variable is the number of years in the past, and we have to subtract this amount from their current ages.

Let $x =$ the number of years in the past

Therefore, x years ago, Jessica was $27 - x$ years old and Melissa was $21 - x$ years old. The relationship x years ago was

Jessica's age $= 2 \times$ Melissa's age

Substituting the past ages, we get the equation for the problem.

$$27 - x = 2(21 - x)$$

Simplifying and solving for x, we get

$$27 - x = 42 - 2x$$
$$x = 15 \text{ years ago}$$

To check, we have to compute the girls' ages 15 years ago. Jessica was $27 - 15 = 12$ years old and Melissa was $21 - 15 = 6$ years old. Clearly, Jessica was twice as old as Melissa.✓

In Chap. 1 we solved problems that make use of many of the basic formulas that arise in mathematics. Those problems merely required that we identify the appropriate formula, substitute known values, and solve for the missing value. Now we will examine situations where those formulas are used as the basis for the equations needed to solve more complex problems. In each problem, we will see that the key to the problem is finding out which of the quantities mentioned are meant to be equal or which quantities are added to get a total amount.

Problems Involving Motion

In problems that involve motion, there are usually one or two objects moving along a straight line at a constant rate (speed) for a period of time. The two objects can be moving toward each other, moving away from each other, or moving in the same direction as in a race. In any situation, there is a total distance traveled by the objects. The governing relationship for these problems is

$$\text{Distance} = \text{rate} \times \text{time}$$

Example 6

Two cars are traveling on the same highway toward each other from towns that are 150 mi apart. One car is traveling at 60 mph, while the other is traveling at 50 mph. If they started at the same time, how much time, to the nearest minute, will have passed at the moment when they pass each other?

Solution 6

In order to use the formula, we have to decide what is given and what we have to find by examining each of the parts. Drawing a picture can help. The car traveling at 60 mph will cover more of the distance than the car traveling at 50 mph.

The picture shows that the total distance traveled is a known amount and that it must be equal to the distances traveled by each car. This is the basis of the model for this problem.

$$\text{Total distance} = \text{distance traveled by faster car}$$
$$+ \text{distance traveled by slower car}$$

We know the rates of each car, and the unknown is the time. Since the cars started at the same time and pass each other at the same time, the time they've both traveled is the same. Therefore, we can use one variable for this time, t.

As before, it is a good idea to state the use of the variable before any work is begun.

$$\text{Let } t = \text{time traveled by both cars}$$

Using the relationship Distance = rate × time for each car, our model becomes

$$150 = 60t + 50t$$
$$150 = 110t$$
$$t = \frac{150}{110}$$
$$t = 1.363636\ldots$$

Remember that this answer must be in hours since the rates were given in miles per *hour*. The problem asks for the answer to be given to the nearest minute. The decimal part of the hour can be converted to minutes by multiplying .37 × 60 = 22.2 minutes and, when rounded, the answer is 1 hour and 22 minutes.

To check the answer, we need to compute the distances traveled by each car. To be accurate, we will use the actual fraction $15/11$ for the time. The faster car traveled $15/11 \times 60 = 900/11$, and the slower car traveled $15/11 \times 50 = 750/11$. By adding these fractions, we find that together they traveled $1650/11 = 150$ mi.✓

Example 7

Allison and Rachel were competing in a race. Allison had a 10-minute head start and was running at 3 km per hour. Rachel ran at 5 km per hour. In how many minutes did Rachel catch up with Allison?

Solution 7

In this problem, the key factor is that at the moment that Rachel catches up to Allison, they will have traveled the same distance. The model for the problem is

Distance traveled by Allison = distance traveled by Rachel

Since we are asked to find Rachel's time, we will represent it by t. Note that the units of t must be hours, since the rates are given in hours. Allison ran 10 minutes more, which will be represented by $t + \frac{1}{6}$, since 10 minutes is one-sixth of an hour.

$$\text{Let } t = \text{Rachel's time}$$

$$t + \frac{1}{6} = \text{Allison's time}$$

Using Distance = rate × time, we have

$$3\left(t + \frac{1}{6}\right) = 5t$$

$$3t + \frac{1}{2} = 5t$$

$$\frac{1}{2} = 2t$$

$$t = \frac{1}{2} \div 2 = \frac{1}{4} \text{ hours} \quad \text{or} \quad t = 15 \text{ minutes}$$

To check, we compute the distance traveled by each girl using $t = \frac{1}{4}$ for Rachel and $t = \frac{1}{4} + \frac{1}{6} = \frac{5}{12}$ for Allison. Rachel's distance was $5 \times \frac{1}{4} = 1.25$ km, and Allison traveled $3 \times \frac{5}{12} = 1.25$ km as well.

Note that we could have converted the rates to kilometers per minute by dividing each rate by 60. This would allow t to represent minutes, and we could use 10 minutes in the equation instead of $\frac{1}{6}$ hour.

$$\left(\frac{3}{60}\right)(t + 10) = \left(\frac{5}{60}\right)t$$

$$\frac{3t}{60} + \frac{30}{60} = \frac{5t}{60}$$

$$3t + 30 = 5t$$

$$30 = 2t$$

$$15 = t$$

The relationship used in these problems can be rewritten to express the rate as a quotient:

$$\text{Rate} = \frac{\text{distance traveled}}{\text{time}}$$

This form of the relationship may be necessary to use in problems when the rates aren't explicitly given.

Problems Involving Work

There are other situations which follow the same principle using a rate. For example, when someone works at a job, we can use

$$\text{Number of jobs completed} = \text{rate of work} \times \text{time}$$

or, equivalently

$$\text{Rate} = \frac{\text{number of jobs completed}}{\text{time}}$$

Example 8

If Joe can paint five rooms in 8 hours and Samantha can paint three rooms in 6 hours, how long will it take them to paint seven rooms working together?

Solution 8

Using the rate formula, we see that Joe's rate of work is $5/8$ rooms per hour and Samantha's rate of work is $1/2$ room per hour.

This problem is actually similar to Example 6, in that they both work for the same amount of time and together they complete the given total amount.

Let $t =$ the time to complete the job together

The model is

Total number of rooms painted

$$= \text{rooms painted by Joe} + \text{rooms painted by Samantha}$$

$$7 = {}^5\!/_8 t + {}^1\!/_2 t = {}^5\!/_8 t + {}^4\!/_8 t$$

$$7 = {}^9\!/_8 t$$

$$56 = 9t$$

$$t = {}^{56}\!/_9 = 6\,{}^2\!/_9 \text{ hours}$$

41

Example 9

It takes Suzanne 10 hours longer than Laurie to take their store's inventory. If it takes $3\frac{3}{4}$ hours for them to do the job together, how long would it take Suzanne to do it alone?

Solution 9

Since Laurie takes less time to complete the job, we will let her time be represented by t. If we can find this time, then the answer to the problem will be $t + 10$.

The number of jobs considered in this problem is 1. Therefore, Laurie's rate is $1/t$ and Suzanne's rate is $1/(t + 10)$. The model for the problem is similar to Example 8, where the total number of jobs completed is the sum of the amount of work done by each girl in the 2 hours.

1 job = amount done by Suzanne + amount done by Allison

Using $\frac{15}{4}$ for $3\frac{3}{4}$, the amount done by Suzanne is $\frac{1}{t+10} \times \frac{15}{4}$, and the amount done by Laurie is $\frac{1}{t} \times \frac{15}{4}$. Substituting into the preceding model, we have

$$1 = \frac{1}{t + 10} \times \frac{15}{4} + \frac{1}{t} \times \frac{15}{4} = \frac{15}{4}\left(\frac{1}{t+10} + \frac{1}{t}\right)$$

$$\frac{1}{{}^{15}/_{4}} = \frac{1}{t + 10} + \frac{1}{t}$$

(Note that this line tells us that the *combined rate is equal to the sum of the individual rates*. If you remember this fact, you can use it as a model when the problem involves only completing 1 job.) Simplify the fraction on the left to get the fractional equation

$$\frac{4}{15} = \frac{1}{t + 10} + \frac{1}{t}$$

Multiplying both sides of the equation by the least common

42

denominator, $15t(t + 10)$, we get

$$4t(t + 10) = 15t + 15(t + 10)$$

Simplifying, we get a quadratic equation $\qquad 4t^2 + 40t = 30t + 150$

Subtracting $30t$ and 150 from both sides gives us $\qquad 4t^2 + 10t - 150 = 0$

Dividing all terms by 2 simplifies the equation to $\qquad 2t^2 + 5t - 75 = 0$

We can now factor and solve
$$(2t + 15)(t - 5) = 0$$
$$2t + 15 = 0 \text{ or } t - 5 = 0$$
$$t = -7.5 \text{ or } t = 5$$

Since the problem requires physical time, we *reject* the negative answer and we see that $t = 5$ is Laurie's time. Therefore, Suzanne can complete the job in 15 hours working alone.

Problems Involving Mixing Quantities

A situation similar to those in the problems above is finding out what occurs when we mix different types of objects. In these problems there is also a rate involved, but it is not a rate involving time, as we will see.

Example 10

At the chocolate factory, John's job is to experiment with mixing different amounts of milk and chocolate to find different flavors. If he mixes 2 quarts (qt) of type A, which contains 15 percent chocolate, with 1 qt of type B, which contains 10 percent chocolate, he creates type C. What percent of type C is chocolate?

Solution 10

In this situation, the rate is the percentage of chocolate. For each type, we have the relationship

Amount of chocolate in the type
$$= \text{percent of chocolate} \times \text{amount of the type}$$

where the percentage is taking the form of the rate as we saw in previous problems.

In type A, we have $0.15 \times 2 = 0.3$ qt of chocolate

In type B, we have $0.10 \times 1 = 0.1$ qt of chocolate

Together, in the mixture type C, we have 0.4 qt of chocolate. Since the mixture has exactly 3 qt of liquid, we calculate the percentage that is chocolate. To do this, we rewrite the relationship (amount of chocolate)/(amount of type) = percentage of chocolate. Since there are $0.1 + 0.3 = 0.4$ qt, the answer is easily found to be $0.4/3 = 0.1333\ldots$ or 13.3%.

Another type of mixture problem involves the price of a mixture of two items where each sells for different individual prices. The underlying formula is also a kind of rate relationship since price is usually given as dollars per pound, cents per pound, or dollars per kilogram. For instance

$$\frac{\text{Cost}}{\text{amount}} = \text{price} \qquad \text{or} \qquad \text{cost} = \text{price} \times \text{amount}$$

Example 11

Nancy's Nut Stand sells cashews for $5/lb and honey-roasted peanuts for $2/lb pound. How many pounds of each type of nut should Nancy use to make 8 lb of a mixture that will sell for $3/lb?

Solution 11

The problem asks us to find the amount of both kinds of nuts. We can solve the problem using only one variable since we know that the total amount must be 8 lb.

Let p = number of pounds of cashews

$8 - p$ = number of pounds of honey-roasted peanuts

The model for the problem is that the total cost for the mixture should be the same as the sum of the costs for each individual kind of nut:

Cost of mixture = cost of cashews

+ cost of honey-roasted peanuts

and using the relationship Cost = price × amount, we have

$$(3)(8) = 5p + 2(8 - p)$$
$$24 = 5p + 16 - 2p$$
$$8 = 3p$$
$$p = \tfrac{8}{3} \quad \text{or} \quad 2\tfrac{2}{3}$$

The mixture should have $2\tfrac{2}{3}$ lb of cashews and $8 - 2\tfrac{2}{3} = 5\tfrac{1}{3}$ lb of honey-roasted peanuts.

Alternative Solution 11

Another way to solve the problem is to use two variables and get two equations to work with. (See the Appendix for a review of how to work with equations with two variables.)

Let x = amount of cashews in mixture

Let y = amount of honey-roasted nuts in mixture

The first equation is the total amount of the mixture

$$x + y = 8$$

The second equation comes from the cost relationship Cost = price × amount, and the model Cost of cashews + cost of honey-roasted peanuts = cost of mixture.

$$5x + 2y = (3)(8) = 24$$

The two equations together form a simultaneous system of equations and can be solved easily by the following process.

Multiply the first equation by 2 and subtract it from the first

$$5x + 2y = 24$$
$$2x + 2y = 16$$
$$3x = 8$$
$$x = \tfrac{8}{3} = 2\tfrac{2}{3}$$

45

Use this value of x in the first equation and solve for y:

$$2\,{}^{2}/_{3} + y = 8$$
$$y = 5\,{}^{1}/_{3}$$

It's okay if you are not familiar with working with two equations. As we saw, all of the problems can be solved with one variable and one equation. However, you will be a better problem solver by having several ways to solve a problem in your arsenal.

Problems Involving Money

One other kind of problem that is easily solved by an algebraic equation involves combinations of money of different denominations. These are problems similar to the previous ones because the amount of money that a coin is worth is a sort of rate. For example, a quarter is worth 25 cents per coin and a dime is worth 10 cents per coin. The amount of money we have is found by the relationship

Amount of money = value of coin × number of coins

Here, the *value of the coin* is taking the form of the rate.

Example 12

Joey has $2.35 in nickels, dimes, and quarters. If he has two less quarters than dimes and one more nickel than dimes, how much of each coin does he have?

Solution 12

The key to the problem is the way in which the numbers of quarters and nickels are related to the number of dimes. We can find the amounts of quarters and nickels if we know the amount of dimes. Therefore, we will assign the variable to this amount.

Let x = number of dimes

Since the number of quarters is two less than this amount, we have

$$x - 2 = \text{number of quarters}$$

Since the number of nickels is one more than the number of dimes, we have

$$x + 1 = \text{number of nickels}$$

The governing relationship is

Total amount of money = amount in quarters
+ amount in dimes + amount in nickels

Using the Amount = rate of coin × number of coins, we have

$$2.35 = .25(x - 2) + 0.10x + 0.05(x + 1)$$

Since we are given the total amount in dollars, we need to use decimals to express the rate of each coin as a decimal fraction of a dollar. If we multiply every term by 100, we can clear the decimals

$$235 = 25(x - 2) + 10x + 5(x + 1)$$

Simplifying and solving for x, we have

$$235 = 25x - 50 + 10x + 5x + 5$$
$$235 = 40x - 45$$
$$280 = 40x$$
$$7 = x$$

Therefore, there are seven dimes, five quarters, and eight nickels. This is easily checked: 70¢ + $1.25 + 40¢ is indeed $2.35.✓

Summarizing what we have done with all our word problems, we saw that we need to follow some basic steps:

Step 1
Read the problem carefully and identify the unknowns.

Step 2
Identify the formula or key concept that is needed in the problem and rewrite the problem, keeping most of the words but using basic arithmetic symbols for operations and equality.

Step 3
Assign a variable to one of the unknowns with a "Let" statement and determine the symbolic representations of all the other knowns with this variable using the relationships in the problem.

Step 4
Substitute the representations of the unknowns into the mathematical statement you created in step 2.

Step 5
Use algebra to solve the equation(s) and determine the value of the unknowns you represented by a variable.

Step 6
Check your answer to see if it really does solve the problem and satisfy all the necessary conditions.

It is worth remembering that in many problems, there are at least two relationships among the unknowns. If only two unknown quantities are asked for, the mathematical model of the problem is sometimes more easily seen when using two variables and creating a system of two equations to solve.

Additional Problems

1. The sum of two numbers is 62. The larger number is as much greater than 34 as the smaller number is greater than 10. What are the two numbers?
2. Separate the number 17 into two parts so that the sum of their squares is equal to the square of one more than the larger part.

3. The difference between two numbers is 12. If 2 is added to 7 times the smaller, the result is the same as when 2 is subtracted from 3 times the larger. Find the numbers.

4. The sum of three consecutive integers is 11 greater than twice the largest of the three integers. What are they?

5. Find three consecutive odd integers whose sum is 75.

6. Find four consecutive even integers such that the sum of the squares of the two largest numbers is 34 more than the sum of the other two.

7. A father is now 5 times as old as his son. In 15 years he will be only twice as old as his son. How old are they now?

8. Sarah's mom is 42 years old, and her aunt is 48 years old. How many years ago was Sarah's aunt 3 times as old as her mom?

9. A cyclist traveling 20 mph leaves ½ hour ahead of a car traveling at 50 mph. If they go in the same direction, how long will it take for the car to pass the cyclist?

10. In the early days of building railroads in the United States, the first transcontinental railroad was started from two points at the ends of the 2000-mi distance over which tracks had to be laid. Two teams at different ends of the country started at the same time. The team in the east moved forward at 2 mi per day over mostly flat land, and the team from the west moved at 1.5 mi per day over mountainous terrain. How long did it take for the teams to meet and the last spike to be nailed in, and how many miles of track did each team lay?

11. On an international flight across the Atlantic Ocean, a certain type of plane travels at a speed that is 220 km/hour more than twice the speed of another. The faster plane can fly 7000 km in the same time that the slower plane can fly 2700 km. At which speeds do these planes fly in miles per hour?

12. Mr. Begly's office has an old copying machine and a new one. The older machine takes 6 hours to make all copies of the financial report, while the new machine takes only 4 hours. If both machines are used, how long will it take for the job to be completed?

13. The candystand at the multiplex has 100 lb of a popular candy selling at $1.20/lb. The manager notices a different candy worth $2.00/lb that isn't selling well. He decides to form a mixture of both types of candy to help clear his inventory of the more expensive type. How many pounds of the expensive candy should he mix with the 100 lb in order to produce a mixture worth $1.75/lb?

14. At her specialty coffee store, Melissa has two kinds of coffee she'd like to mix to make a blend. One sells for $1.60/lb and the other, $2.10/lb. How many pounds of each kind must she use to make

75 lb of mixed coffee that she can sell for $1.90/lb and make the same profit?

15. Joey has $5.05 in 34 coins on his dresser. If the coins are quarters and dimes, how many of each kind are there?

16. The triplets John, Jen, and Joey gave their father all their coins toward a present for their mother. John had only quarters to give. Jen had only dimes and gave 4 times as many coins as John gave. Joey had only nickels and gave seven more than twice the number of coins that Jen gave. If the total amount of the contribution was $22.40, how much did each child contribute?

17. Mrs. Stone likes to diversify her investments. She invested part of $5000 in one fund that had a return of 9% interest and the rest into a different fund that earned 11%. If her total annual income from these investments was $487, how much does she invest at each rate?

18. Three people each received part of $139 so that the first received twice what the second received and the third's portion exceeded the sum of the other two by one dollar. How much money did each person receive?

19. A farmhand was hired for a 60-day period and would live on the ranch. For each day that he worked he would receive $45, but for each day that he did not work he would pay $10 for his room and board. At the end of the 60 days, he received $2260. How many days did he work?

Solutions to Additional Problems

1. Using one variable, we have the following. Let x = the smaller number; therefore, $62 - x$ is the larger number. In Prob. 1, the expression "as much greater than" indicates that we want the difference between the numbers.

$$\text{The larger number} - 34 = \text{the smaller number} - 10$$
$$(62 - x) - 34 = x - 10$$
$$28 - x = x - 10$$
$$38 = 2x$$
$$x = 19$$

The smaller number is 19, and the larger number is $62 - 19 = 43$.

Check: $43 - 34 = 9$ and $19 - 10 = 9.\checkmark$ An alternative solution using two variables is as follows. Let $x =$ smaller number and let $y =$ larger number. The equations would be $x + y = 62$ and $y - 34 = x - 10$. The second equation could be rewritten as $-x + y = 24$. Adding the first equation to this new form of the second equation gives us $2y = 86$ and $y = 43$. Substituting into the first equation, we have $x + 43 = 62$ and $x = 19$.

2. Both "parts" of the number 17 can be represented using one variable. We should use the single variable to represent the larger part, since another piece of the puzzle relates to it. Let $x =$ the larger part and $17 - x =$ smaller part and $x + 1 =$ one more than the larger part. Recognizing the *sum of their squares* to mean *larger part^2 + smaller part^2*, we have the quadratic equation

$$x^2 + (17 - x)^2 = (x + 1)^2$$

Squaring the binomials, we have

$$x^2 + 289 - 34x + x^2 = x^2 + 2x + 1$$

Combining terms gives us

$$2x^2 - 34x + 289 = x^2 + 2x + 1$$

Subtracting terms on the right from both sides of the equation produces

$$x^2 - 36x + 288 = 0$$

We can factor this and solve

$$(x - 12)(x - 24) = 0$$
$$x - 12 = 0 \quad \text{or} \quad x - 24 = 0$$
$$x = 12 \quad \text{or} \quad x = 24$$

Since the larger part of 17 must be less than 17, we reject the answer of 24. The larger part is 12 and the smaller part is 5.

Check: $12 + 5 = 17$. $12^2 + 5^2 = 144 + 25 = 169$ and $(12 + 1)^2 = 13^2 = 169.\checkmark$

3. The problem is best understood and modeled when two variables are used for larger and smaller numbers. The model is

$$\text{Larger number} - \text{smaller number} = 12$$

$$7 \times \text{smaller} + 2 = 3 \times \text{larger} - 2$$

Let $x = $ larger number and let $y = $ smaller number. The two equations are $x - y = 12$ and $7y + 2 = 3x - 2$. We can rewrite the second equation as $-3x + 7y = -4$, and we will be able to eliminate one of the variables by adding equations if we multiply every term in the first equation by 7: $7x - 7y = 84$. Adding the equations gives us $4x = 80$ and $x = 20$. Using the original first equation to find y, we have $20 - y = 12$ and $y = 8$. The numbers are 20 and 8.

Check: The difference of the numbers is $20 - 8 = 12$. The other relationships are

$$7 \times 8 + 2 = 56 + 2 = 58$$

$$3 \times 20 - 2 = 60 - 2 = 58\checkmark$$

4. *Consecutive* integers imply that the three numbers are one apart. Therefore

$$\text{Let } x = \text{first (smallest) number}$$
$$x + 1 = \text{second (middle) number}$$
$$x + 2 = \text{third (largest) number}$$

The model is

$$\text{First number} + \text{second number} + \text{third number}$$

$$= 2 \times \text{third number} + 11$$

(Be sure not to confuse the phrase "is 11 greater than" with "11 is greater than," which translates into an inequality "11 >"!) Then $x + (x + 1) + (x + 2) = 2(x + 2) + 11$. This linear equation can be solved in the usual way. (At this point, see if you can identify what steps are being taken.)

$$3x + 3 = 2x + 4 + 11$$

$$3x + 3 = 2x + 15$$

$$x = 12$$

The numbers are 12, 13, and 14.

Check: 12, 13, and 14 are *consecutive* integers. $12 + 13 + 14 = 39$ and $2 \times 14 + 11 = 28 + 11 = 39.\checkmark$

5. By definition, an *odd number* is an integer (fractions are not considered to be odd or even). To be consecutive odd numbers the numbers must differ by 2. Therefore

    Let x = first (smallest) odd number

    $x + 2$ = second (middle) odd number

    $x + 4$ = third (largest) odd number

The model is simply

First odd number + second odd number

$+$ third odd number $= 75$

The equation is therefore

$$x + (x + 2) + (x + 4) = 75$$
$$3x + 6 = 75$$
$$3x = 69$$
$$x = 23$$

The three consecutive odd numbers are 23, 25, and 27.

Check: $23 + 25 + 27 = 75.\checkmark$

6. Consecutive even integers must be 2 apart. Therefore,

    Let x = first (smallest) even integer

    $x + 2$ = second even integer

    $x + 4$ = third even integer

    $x + 6$ = fourth (largest) even integer

The model for the problem is

(Third integer)^2 + (fourth integer)^2 = first integer

$+$ second integer $+ 36$

The equation is therefore

$$(x + 4)^2 + (x + 6)^2 = x + (x + 2) + 34$$
$$x^2 + 8x + 16 + x^2 + 12x + 36 = 2x + 36$$
$$2x^2 + 20x + 52 = 2x + 36$$
$$2x^2 + 18x + 16 = 0$$
$$x^2 + 9x + 8 = 0$$
$$(x + 1)(x + 8) = 0$$
$$x + 1 = 0 \quad \text{or} \quad x + 8 = 0$$
$$x = -1 \quad \text{or} \quad x = -8$$

Since we want even integers, we reject $x = -1$ and our smallest even integer is -8. The four even integers are -8, -6, -4, and -2.

Check: $(-4)^2 + (-2)^2 = 16 + 4 = 20$ and $-8 + -6 + 34 = 20.\checkmark$

Don't be disturbed by having negative numbers as answers. There is no reason to assume that all number problems involve only positive integers!

7. The two relationships are clearly stated in the problem. One relationship can serve as the basis for assigning a variable. Since the father's age is given in terms of the son's age

$$\text{Let } x = \text{son's age now}$$
$$5x = \text{father's age now}$$

The second relationship provides us with the model

$$\text{Father's age now} + 15 = 2 \times (\text{son's age now} + 15)$$
$$5x + 15 = 2(x + 15)$$
$$5x + 15 = 2x + 30$$
$$3x = 15$$
$$x = 5$$

The son is now 5 and the father is $5 \times 5 = 25$.

Check: In 15 years the son will be 20 and the father will be 40. $40 = 2 \times 20.\checkmark$

8. The unknown in this problem is the number of years in the past from now. The model for the problem is

Sarah's aunt's age in the past

$= 3 \times$ Sarah's mom's age in the past

Let $x =$ the number of years from then until now. Sarah's aunt's age was $48 - x$. Sarah's mom's age was $42 - x$. The equation is therefore

$$48 - x = 3(42 - x)$$
$$48 - x = 126 - 3x$$

To both sides we add $3x$ and subtract 48 to get the simple equation $2x = 78$. Dividing both sides by 2 gives us $x = 39$. Therefore, 39 years ago, Sarah's aunt was 3 times as old as Sarah's mom.

Check: 39 years ago, Sarah's aunt was $48 - 39 = 9$ and Sarah's mom was $42 - 39 = 3$. $9 = 3 \times 3$. ✓

9. The "hidden" concept here is that at the moment when the car passes the cyclist, they have traveled the same distance. The model for the problem will be

Distance traveled by car $=$ distance traveled by cyclist

For each distance, we make use of the formula Distance $=$ rate $\times$ time. Since time is the unknown and the cyclist travels 0.5 hour longer than the car, we have

Let $t =$ time traveled by car

$t + 0.5 =$ time traveled by cyclist

The equation we have from the model is

$$20 \times (t + 0.5) = 50t$$
$$20t + 10 = 50t$$
$$10 = 30t$$
$$t = 1/3 \text{ hour or 20 minutes}$$

Check: The cyclist travels for $1/3 + 1/2 = 5/6$ hours and covers a distance of $20 \times 5/6 = 100/6$ or $50/6$ mi. The car travels for $1/3$ hour and covers a distance of $50 \times 1/3 = 50/3$ mi. ✓

10. This is really a "motion" problem when thinking of the teams as very slow moving vehicles. The model for the problem is based on the fact that the total distance is equal to the sum of the distances covered by both teams or

Distance covered by team from east

$+$ distance covered by team from west $= 2000$

The "hidden" fact here is that when two objects are moving along the same path toward each other, then at the point when they meet, they have spent the same amount of time traveling! Since the rate is given in miles per *day*, we let $t =$ the number of days spent by both teams. Then, using Distance $=$ rate $\times$ time for each team's distance, we have

$$2\,\text{mi/day} \times t + 1.5\,\text{mi/day} \times t = 2000$$

$$2t + 1.5t = 2000$$

$$3.5t = 2000$$

$$t = 2000 \div 3.5 = 571.43\,\text{days}$$

or approximately $571.43/365 = 1.56$ or $1\frac{1}{2}$ years. The team from the east laid approximately $2 \times 571.43 = 1142.8$ or 1143 mi of track. The team from the west laid approximately $1.5 \times 571.43 = 857.1$ or 857 miles of track.

Check: Since the problem called for the amount of track laid, the only necessary check is that $1143\,\text{mi} + 857\,\text{mi} = 2000\,\text{mi}$.

11. Notice that the problem specifies that we give our answer in miles per hour. This adds only one more step in the solution as we can find the rates in kilometers per hour and convert to miles per hour at the end. Working in the metric system does not change the way in which we solve the problem.

There are clearly two relationships given in the problem. The first is straightforward and allows us to use one variable to represent both rates.

$$\text{Let } r = \text{rate of slower plane}$$

$$2r + 220 = \text{rate of faster plane}$$

The problem gives us the distances for the same amount of time traveled by both planes. From Distance = rate × time, we have the alternative form of Time = distance/rate, and we can derive the model

$$\frac{\text{Distance of fast plane}}{\text{Rate of fast plane}} = \frac{\text{distance of slow plane}}{\text{rate of slow plane}}$$

and the equation

$$\frac{7000}{2r + 220} = \frac{2700}{r}$$

Cross-multiplying, we have

$$7000r = 2700(2r + 220)$$

Simplifying and solving, we obtain

$$7000r = 5400r + 594{,}000$$

$$1600r = 594{,}000$$

$$r = 371.25$$

The slower plane is traveling at 371.25 km/hour and the faster plane is traveling at 962.50 km/hour. Multiplying each of these rates by the conversion factor of 0.62 mi/km, we have that the slower plane is traveling at approximately 230 mph and the faster plane can travel at 597 mph.

Check: For the check, we will use the metric answers. The slow plane travels 2700 km in 2700 ÷ 371.25 = 7.2727 ... hours. In this time, the fast plane travels 962.5 × 7.2727... = 7000 km.✓

12. The model here states that if the machines are working together for the same amount of time, each will do a fraction of the job.

Therefore

$$1 \text{ job} = \text{fraction done by old machine}$$
$$+ \text{ fraction done by new machine}$$

The rate of each machine can be found from

$$\text{Rate} = \frac{\text{number of jobs completed}}{\text{time}}$$

The rate of the older machine is $1/6$ job/hour, and the rate of the new machine is $1/4$ job/hour. Let $t =$ the number of hours the machines work together.

Using the Number of jobs completed = rate of work × time for each machine, we have

$$1 = 1/6 t + 1/4 t$$
$$1 = 5/12 t$$
$$t = 12/5 \text{ or } 2.4 \text{ hours}$$

Check: In 2.4 hours the older machine has done $2.4/6 = 0.4$ of the job and the newer machine has done $2.4/4 = 0.6$ of the job; thus $0.4 + 0.6 = 1$ job.✓

13. The underlying concept here is that the amount of money earned from selling the mixture at \$1.75 per pound should equal the total of having sold each of the two candies at their normal price or

$$\text{Dollar amount from mixture}$$
$$= \text{dollar amount from cheaper candy}$$
$$+ \text{dollar amount from expensive candy}$$

We use the formula Dollar amount = price × amount sold. Let $x =$ the number of pounds of expensive candy and we have that $x + 100$ be the amount of the mixture. The equation from the model is $1.75 (x + 100) = 1.20 \times 100 + 2.00x$. Simplifying the equation and multiplying every term by 100 gives us

$$175x + 17{,}500 = 12{,}000 + 200x$$
$$5500 = 25x$$
$$x = 220 \text{ lb}$$

Check: When all candy is sold, the mixture would bring in $1.75 \times 320 = \$560$; specifically, 100 lb of the cheaper candy would bring in $1.20 \times 100 = \$120$, and 220 lb of the expensive candy would bring in $2.00 \times 220 = \$440$. $\$120 + \$440 = \$560.\checkmark$

14. The phrase "make the same profit" indicates that Melissa is seeking to receive the same amount of money from the blend as she would if she sold the same amounts that are in the mixture separately. The model is therefore

> Dollar amount received from blend
>
> $=$ dollar amount from cheaper coffee
>
> $+$ dollar amount from the expensive coffee

The two amounts of coffee can be represented by the same variable easily.

> Let $x =$ number of pounds of cheaper coffee in blend
>
> $75 - x =$ number of pounds of expensive coffee in blend

Each amount is found by the formula Dollar amount $=$ price $\times$ amount used. Therefore, our equation is

$$1.90 \times 75 = 1.60x + 2.10(75 - x)$$
$$142.50 = 1.60x + 157.5 - 2.10x$$
$$-15 = -0.5x$$
$$x = 30$$

Melissa needs to use 30 lb of the cheaper coffee and 45 lb of the more expensive coffee.

Check: The amount from the blend is $\$142.50$, specifically, $\$1.60 \times 30 = \$48.00$ and $\$2.10 \times 45 = \$94.50$. $\$48.00 + \$94.50 = 142.50.\checkmark$

15. The two relationships in this problem are

> Total number of coins $= 34$
>
> Value of the coins $= \$5.05$

Let d = the number of dimes and let q = the number of quarters. The first relationship gives us the equation $d + q = 34$. The second relationship uses the following concepts.

$$\text{Amount of money} = \text{value of coin} \times \text{number of coins}$$

$$\text{Total amount of money} = \text{amount in quarters}$$
$$+ \text{amount in dimes}$$

The second equation is therefore $0.10d + 0.25q = 5.05$. To get both equations into a form in which we can add or subtract them to eliminate one of the variables, we multiply the first equation by 25 and the second equation by 100 to get

$$25d + 25q = 850$$
$$10d + 25q = 505$$

and we subtract to get $15d = 345$. Therefore, $d = 23$, and from the original first equation

$$23 + q = 34 \qquad \text{or} \qquad q = 11$$

Check: 23 dimes + 11 quarters = 34 coins and $\$.10 \times 23 + \$.25 \times 11 = \$2.30 + \$2.75 = \$5.05.\checkmark$ (Try solving the problem, using d for the number of dimes and $34 - d$ for the number of quarters.)

16. The model for the problem is given by

$$\text{Amount from John's quarters} + \text{amount from Jen's dimes}$$
$$+ \text{amount from Joey's nickels} = \$22.40$$

Each amount is calculated by Amount of money = value of coin × number of coins. We should try to express all three numbers of coins with the same variable instead of using three different variables. To do this, we need to study the given relationships carefully. Let x = number of coins (quarters) John gave. Jen gave 4 times as many coins as John, so

$$4x = \text{number of coins (dimes) Jen gave}$$

Joey gave 2 times the number of coins Jen gave + 7, so

$$8x + 7 = \text{number of coins (nickels) Joey gave}$$

Using these representations in the model, we have

$$0.25x + 0.10 \times 4x + 0.05(8x + 7) = 22.40$$

Simplifying and multiplying every term by 100, we have

$$25x + 40x + 40x + 35 = 2240$$
$$105x + 35 = 2240$$
$$105x = 2205$$
$$x = 21$$

Using this in our "Let" statements, we deduce that John gave 21 quarters, Jen gave $4 \times 21 = 84$ dimes, and Joey gave $8 \times 21 + 7 = 175$ nickels.

Check: 84 is $4 \times 21 \checkmark$ and 175 is $2 \times 84 + 7. \checkmark$ Also

$$\$0.25 \times 21 + \$0.10 \times 84 + \$0.05 \times 175$$
$$= \$5.25 + \$8.40 + \$8.75 = \$22.40\checkmark$$

17. This problem is similar to those we've looked at in that the underlying model is

$$\text{Total interest earned} = \text{interest earned from 9\% fund}$$
$$+ \text{interest earned from 11\% fund}$$

We use the formula

$$\text{Interest earned} = \text{principal} \times \text{rate of interest}$$
$$\times \text{time invested}$$

and note that the time over which this interest was earned was 1 year. The unknown in the problem is the amount invested in each fund. Since the total amount is $5000, we let x = the amount invested at 9% and $5000 - x$ = the amount invested at 11 percent. The model gives us the equation

$$487 = 0.09x + 0.11(5000 - x)$$
$$48,700 = 9x + 55,000 - 11x$$
$$48,700 = 55,000 - 2x$$
$$2x = \$6300$$
$$x = \$3150$$

Mrs. Stone invested $3150 in the 9% fund and the balance, $5000 - \$3150 = \$1850$ in the 11 percent fund.

Check: The interest from the 9 percent fund was $0.09 \times \$3150 = \$283.50$, and the interest from the 11 percent fund was $0.11 \times \$1850 = \$203.50$; then $\$283.50 + \$203.50 = \$487.00.\checkmark$

18. It would become very complicated to try to use three variables to represent the different amounts that each person received. This forces us to try to use only one variable to model the problem. The situation is clearest when we use the second person's amount as the single variable since the first is easily determined when we know this amount. Let x = amount received by second person. Clearly, $2x$ = amount received by first person.

The third person must have received the remaining portion of the $139 or $139 - (x + 2x)$.

$$139 - 3x = \text{amount received by third person}$$

Note that the phrase "exceeds by some amount" can be translated as "is that amount more than." The model is, therefore

Third person's portion = first person's portion
$$\qquad\qquad + \text{second person's portion} + \$1$$
$$139 - 3x = 2x + x + 1$$
$$139 - 3x = 3x + 1$$
$$138 = 6x$$
$$x = 23$$

62

The second person received $23, the first person received $46, and the third person received $139 − $69 = $70.

Check: $23 + $46 + $70 = $139; then $23 + $46 + $1 = $70.✓

19. This problem doesn't belong to any category that was mentioned in this chapter, but it is still basically a "rate × amount" problem and is easily solved with some algebra. The rates are $45 earned per day and $10 paid per day. Having two variables makes the modeling easy. (See the Appendix for a review of how to work with equations having two variables.)

Let x = number of days that the farmhand worked

Let y = number of days that the farmhand did not work

We know that the total number of days is 60. Therefore, the first equation is

$$x + y = 60$$

The basic concept is that

Amount earned from working

$$- \text{amount paid for room and board} = \text{amount received}$$

The second equation is, therefore, $45x − 10y = 2260$. Multiplying every term in the first equation by 10 and adding this to the second equation gives

$$10x + 10y = 600$$
$$45x - 10y = 2260$$

Adding the two equations gives us $55x = 2860$, and we find $x = 52$. The farmhand worked for 52 days and did not work for 8 days.

Check: $52 + 8 = 60$; then $52 × $45 − 8 × $10 = $2340 − $80 = $2260.✓

Word Problems Involving Ratio, Proportion, and Percentage

Many situations presented in word problems involve comparison of the sizes of groups and objects. The answer sought may not be the number of objects, but the fraction that indicates how much of the larger group is the smaller or how two groups are related to the whole. These fractions are referred to as *ratios*.

Sometimes when ratios are given, they are given by the statement *a:b*. This is equivalent to the fraction *a/b*. The fraction allows us to use arithmetic in the usual way to solve problems.

Example 1

In a math class, there are 32 students. If the ratio of boys to girls is 3:5, how many boys and how many girls are there in the class?

Solution 1

One way to understand this problem is to restate the ratio in the following way. For every three boys there are five girls. With this we can create a visual representation of the situation using B for a boy and G for a girl. Writing groups of BBBGGGGG and keeping track of how many we have in a cumulative way gives us:

Groups	Running count
BBBGGGGG	8
BBBGGGGG	16
BBBGGGGG	24
BBBGGGGG	32

Adding the Bs and Gs tells us that there are 12 boys and 20 girls. The check for this answer is to create the fraction of boys/girls and see if it is equal to $3/5$.

Check: $12/20 = 0.6 = 3/5.$ ✓

Alternative Solution 1

The method described above was quick, but it would be tedious if the group were much larger than 32. The problem really involves finding the number of groups and applying a rate. The ratio 3 boys:5 girls can be translated into two rates:3 boys/group and 5 girls/group. The model for the situation is

Class size = number of boys + number of girls

where each number is found by

Rate of boys or girls per group × the number of groups

(In Chap. 2, there are many problems using similar models.)
   If we think of the number of groups as the unknown, we can solve the problem algebraically by assigning a variable to this and modeling the situation with an equation.

Let x = the number of groups

Number of boys = $3x$

Number of girls = $5x$

Using these quantities in the model, we have the equation

$$3x + 5x = 32$$
$$8x = 32$$
$$x = 4$$

There are four groups, each consisting of 3 boys and 5 girls. Therefore, there are $3 \times 4 = 12$ boys and $5 \times 4 = 20$ girls.

Example 2

An advertisement for toothpaste claims that 4 out of 5 dentists recommend a certain brand. A consumer group wanted to check the accuracy of this claim and surveyed 180 dentists. How many dentists would they expect to not recommend this brand, if the advertisement's claim is accurate?

Solution 2

The phrase "4 out of 5" indicates that the ratio of those who recommend to those who do not recommend is 4:1. Using the picture scheme R for recommend and D for doesn't recommend, we would have to list and count groups of RRRRD. Rather than listing groups, we easily realize that there has to be $180/5 = 36$ groups and, therefore, there would be 36 Ds if the claim were accurate.

The more structured algebraic approach would be

Let x = the number
of groups

Number of those who recommend = $4x$

Number of those who do not recommend = x

$$4x + x = 180$$
$$5x = 180$$
$$x = 36$$

There would be $4 \times 36 = 144$ dentists who recommend the brand and $1 \times 36 = 36$ who do not.

Ratios can appear in many of the types of problems we saw in the previous chapter.

Number Problems

Example 3

Two positive numbers are in the ratio of 5:8. Find the numbers, assuming that the square of the smaller is 15 less than 10 times the larger.

Solution 3

If we think of multiples of 5 and 8 as "groups," we can use a solution similar to that for Probs. 1 and 2; specifically, we can list pairs of multiples and check the conditions for each pair.

Multiple	Smaller	Larger	Smaller²	10 × larger−15
1	5	8	25	$80 - 15 = 65$
2	10	16	100	$160 - 15 = 145$
3	15	24	225	$240 - 15 = 225\checkmark$

As before, this is an easy solution because the actual numbers came up early in the list. If the actual numbers were larger, listing them in this fashion would be tedious.

The algebraic approach would be

$$\text{Let } x = \text{"multiple" needed}$$

$$5x = \text{smaller number}$$

$$8x = \text{larger number}$$

The equation would come from the model

$$\text{Smaller}^2 = 10 \times \text{larger} - 15$$

and we would have the quadratic equation:

$$(5x)^2 = 10(8x) - 15$$

$$25x^2 = 80x - 15$$

$$25x^2 - 80x + 15 = 0$$

Dividing every term in the equation by 5 gives us the simpler quadratic

$$5x^2 - 16x + 3 = 0$$

$$(5x - 1)(x - 3) = 0$$

$$5x - 1 = 0 \text{ or } x - 3 = 0$$

$$x = \tfrac{1}{5} \quad \text{or} \quad x = 3$$

Since x is a multiple, it must be a whole number. Therefore, we reject the answer of $\frac{1}{5}$. Our numbers would be the third multiples of 5 and 8 or 15 and 24.

Check: $15^2 = 225$; then $10 \times 24 - 15 = 240 - 15 = 225.\checkmark$

Age Problems

Example 4

Jason is Chris' older cousin. The ratio of their ages is 9:5. In 5 years, Chris' age will be 4 more than half of Jason's age. How old are the boys now?

Solution 4

The ratio indicates that the current ages of the boys are multiples of 5 and 9. Making a table as follows will lead to the answer quickly.

Now		In 5 years		$\frac{1}{2}$ Jason's age	Chris' age $- \frac{1}{2}$ of Jason's age
Chris	Jason	Chris	Jason		
5	9	11	15	$7\frac{1}{2}$	$-3\frac{1}{2}$
10	18	15	23	$11\frac{1}{2}$	$+3\frac{1}{2}$
15	27	20	32	16	$+4$ $\checkmark$

Alternative Solution 4

Using algebra

$$\text{Let } x = \text{the multiple}$$
$$\text{Chris' current age} = 5x$$
$$\text{Jason's current age} = 9x$$

The model is

$$\text{Chris' age} + 5 = \frac{1}{2} \text{ (Jason's age} + 5) + 4$$

Substitution gives

$$5x + 5 = \frac{1}{2}(9x + 5) + 4$$
$$5x + 5 = 4.5x + 2.5 + 4$$

$$5x + 5 = 4.5x + 6.5$$
$$0.5x = 1.5$$
$$x = 3$$

Chris' age is $5 \times 3 = 15$, and Jason's age is $9 \times 3 = 18$.

Check: In 5 years their ages would be 20 and 32. Specifically, $20 = \frac{1}{2} \times 32 + 4 = 16 + 4 = 20.$✓

Perimeter

Example 5

The perimeter of a triangular plot of land is 273 m. A surveyor is making a map of the land and notes that the ratio of the sides is 2:9:10. What is the actual length of each side of the plot?

Solution 5

In solving this problem, the use of a table would be a lengthy task. An algebraic approach gets to the heart of the problem quickly. (Make sure to draw a diagram that reflects the ratios to help visualize the problem accurately.)

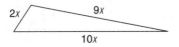

Let x be the multiple.
The sides of the triangular plot are $2x$, $9x$, and $10x$.
The model is

$$\text{Sum of all sides} = 273$$

We have the equation

$$2x + 9x + 10x = 273$$
$$21x = 273$$
$$x = 13$$

The sides are therefore $2 \times 13 = 26$ m, $9 \times 13 = 117$ m, and $10 \times 13 = 130$ m.

Check: $26 + 117 + 130 = 273.\checkmark$

Mixture

Example 6

Carole's candystand sells a popular mixture of two of her candies which sell for $2.70/lb and $1.95/lb. The ratio of the expensive candy to the cheaper candy in the mixture is 3:5. If Carole sold her entire supply of the mixed candy during the coming month, she would have revenue of $928.20, which would be the same as if she sold each candy separately. How many pounds of candy does she have in supply, and at what price does she sell it for?

Solution 6

This seems like a more complicated problem than the previous ones in that more information is given. The key to the problem is that the revenue would be the same as if she sold each type of candy separately. Sometimes when a problem seems complicated, we should try to use a simple number and work with the information in the problem to help get a better understanding of what is occurring. For example, since the ratio of the candies in the mixture is 3:5, let's assume that 8 lb was sold. In other words, we might ask, "What would be the revenue if 3 lb of the expensive candy and 5 lb of the cheaper candy were sold?"

Clearly, the answer to this question is $3 \times \$2.70 + 5 \times \$1.95 = \$17.85$. The price of the mixture would be $17.85 \div 8$ lb, giving us $2.23/lb (rounded).

We easily see that 8 lb is not the answer we are looking for, but it helped us model the problem. The model is

Number of pounds of expensive candy × $2.70 +
number of pounds of cheaper candy × $1.95 = $928.20

Using algebra:

$$\text{Let } x = \text{the multiple}$$
$$\text{Number of pounds of expensive candy} = 3x$$
$$\text{Number of pounds of cheaper candy} = 5x$$
$$\text{Total number of pounds sold} = 8x$$

The model gives us the equation

$$3x \times 2.70 + 5x \times 1.95 = 928.20$$
$$17.85x = 928.20$$
$$x = 52$$

Therefore, $8 \times 52 = 416$ lb of the mixture is in supply and sells for $928.20 \div 416 = \$2.23/\text{lb}$. (Note that this is the same price as when 8 lb is sold. This would be the price for any amount sold, since we are maintaining the same ratio in all quantities of the mixture.)

Check: The mixture contains $3 \times 52 = 156$ lb of the expensive candy, which would have a revenue of $156 \times \$2.70 = \$421.20$, and $5 \times 52 = 260$ lb of the cheaper candy, which would have a revenue of $260 \times \$1.95 = \$507.00$. The total revenue is $\$421.20 + \$507.00 = \$928.20.\checkmark$

Proportion

Word problems sometimes involve comparing two situations that are numerically or geometrically similar. This often requires a scaling of the quantities in a *proportional* manner. For example, a numerically similar situation could be working with a recipe that lists quantities required to make four servings needs to be scaled to make more servings when more than four are required. A geometrically similar situation could be working with two rectangles that are of different sizes but of the same length:width ratio.

When dealing with such situations, we create a mathematical statement indicating that the ratios are the same. This is called a *proportion* and would appear in ratio form as $a{:}b = c{:}d$ or in fraction form as $a/b = c/d$.

From the ratio form, we call a and d the *extremes* while b and c are referred to as the *means*. From the fraction form, we see that if we multiply both sides of the equation by bd, we arrive at $ad = bc$. This is sometimes referred to as *cross-multiplying in order to solve a proportion*. The more mathematical phrase is: *In a proportion, the product of the means is equal to the product of the extremes.* This will be the governing principle in solving the following problems.

Measurement of Quantities

Example 7

Among the ingredients in a recipe for a sweet dessert are $2\frac{1}{2}$ cups of regular sugar and $\frac{1}{4}$ cup of brown sugar. The directions indicate that this will serve 8 people. If there will be 12 people to serve, how much of each type of sugar will be needed?

Solution 7

The underlying assumption in the problem is that the recipe needs to be scaled proportionally when seeking to make more servings. The first step is to decide which two of the three quantities given in the original recipe are necessary for the ratios that will appear in the proportion: $2\frac{1}{2}$, $\frac{1}{4}$, or 8. It is helpful in solving problems of this kind to rephrase the question so that the problem is better understood: "What happens when the number of people changes to 12?" This indicates that our ratios and proportion must involve 8 and 12.

This also helps us see that we will need to solve two different proportions: one for regular (refined white) sugar and the other for brown sugar. The model for the problem is, therefore

(a) The amount of regular sugar for 8 people *is proportional to* the amount of regular sugar for 12 people.
(b) The amount of brown sugar for 8 people *is proportional to* the amount of brown sugar for 12 people.

Using (a), we can assign a variable and proceed algebraically. Let x be the amount of regular sugar for 12 people. The model gives us the equation

$$\frac{2\frac{1}{2}}{8} = \frac{x}{12}$$

Cross-multiplying, we have $8x = 30$ and $x = 30/8 = 3\frac{3}{4}$ cups of regular sugar. Using (*b*), we can solve in a similar fashion. Let *y* be the amount of brown sugar for 12 people

$$\frac{\frac{1}{4}}{8} = \frac{y}{12}$$

Cross-multiplying, we have $8x = 3$ and $y = \frac{3}{8}$ cup of brown sugar.

Check: One way to check the problem is to see if the ratios of the original amounts to the new amounts is the same as 12:8 or $\frac{3}{2}$. For regular sugar, $3\frac{3}{4} \div 2\frac{1}{2} = \frac{15}{4} \div \frac{5}{2} = \frac{15}{4} \times \frac{2}{5} = \frac{30}{20} = \frac{3}{2}.\checkmark$

For brown sugar, $\frac{3}{8} \div \frac{1}{4} = \frac{3}{8} \times \frac{4}{1} = \frac{12}{8} = \frac{3}{2}.\checkmark$

Measurement of Similar Figures

Example 8

On a blueprint for a new office building, the rectangular conference room measures 4.5 in by 12 in. The shorter wall of the actual room measures 15 ft. How much carpeting will be needed to cover the floor of the actual room?

Solution 8

The underlying assumption is that a blueprint is a *scaled* drawing of the room; that is, the sides of the room on the blueprint are proportional to the actual sizes or the ratios of the corresponding sides are equal. The model for the problem is (length on blueprint)/(actual length) = (width on blueprint)/(actual width).

Once we find the actual width, we need to compute the area of the floor of the room. Letting *w* be the actual width, we have $4.5/15 = 12/w$. Cross-multiplying gives us $4.5w = 180$. Therefore, $w = 180/4.5 = 40$ ft. The area is $15 \times 40 = 600$ ft^2.

In Chap. 1, we were concerned about how to keep track of units of measure. Note that in computation of *w*, we actually have the following calculation, including units: $w = (12 \text{ in} \times 15 \text{ ft})/4.5 \text{ in} = 40 \text{ ft}$, which shows that we have made the correct calculation as it gives us our answer in feet when we cancel units.

We see that something interesting happened when we compare the areas of the blueprint drawing to the actual drawing. The area of the room on the blueprint is $4.5 \text{ in} \times 12 \text{ in} = 54 \text{ in}^2$, and the area of the actual room is $600 \text{ ft}^2 = 600 \text{ ft}^2 \times 144 \text{ in}^2/1 \text{ ft}^2 = 86{,}400 \text{ in}^2$. The ratios of the two areas is $86{,}400 \text{ in}^2/54 \text{ in}^2 = 1600$. The length of the actual room is $15 \text{ feet} = 15 \text{ ft} \times 12 \text{ in/ft} = 180 \text{ in}$, and the ratio of the lengths is $180 \text{ in}/4.5 \text{ in} = 40$. What we notice is that $1600 = 40^2$. In general, *the ratio of the areas is the square of the ratios of the lengths*. This fact is useful to remember.

Travel

From the formula for motion

$$\text{Distance} = \text{rate} \times \text{time}$$

we have the following equivalent formula.

$$\text{Time} = \frac{\text{distance}}{\text{rate}}$$

In a problem where the time taken is the same for similar modes of transportation, we can set up a proportion between the distances and rates.

Example 9

Test drives of a new automobile indicated that while traveling at two speeds, one of which was 20 mph greater than the other, the higher speed covered a test distance of 1000 ft while the lower speed covered a distance of 800 ft in the same time. What were the tested speeds, and what was the test time to the nearest second?

Solution 9

Since the test times were the same, we can use the ratio of distance/rate for each situation and create a proportion to model the problem:

$$\frac{\text{Distance of slower test}}{\text{Slower rate}} = \frac{\text{distance of faster test}}{\text{faster rate}}$$

Before beginning, however, we need to convert the distances into miles since the rates are given in miles per hour.

$$\frac{1000 \text{ ft}}{1} \times \frac{1 \text{ mi}}{5280 \text{ ft}} = 0.189 = 0.19 \text{ mi}$$

$$\frac{800 \text{ ft}}{1} \times \frac{1 \text{ mi}}{5280 \text{ ft}} = 0.152 = 0.15 \text{ mi}$$

Let r be the slower rate and $r + 20$ be the faster rate. The model gives us the equation

$$\frac{0.15}{r} = \frac{0.19}{r + 20}$$

Cross-multiplying, we have $0.15(r + 20) = 0.19r$.

Multiplying by 100, we have $15(r + 20) = 19r$, and we can solve this equation in the usual way.

$$15r + 300 = 19r$$

$$300 = 4r$$

$$r = 75 \text{ mph}$$

The slower tested speed was 75 mph, while the faster tested speed was 95 mph. Calculating the time tested actually gives us the check on the problem since both distances and speeds should produce the same time.

$$\textit{Time of slower test} = \frac{\text{distance}}{\text{rate}} = \frac{0.15}{75} = 0.002 \text{ hour}$$

$$\textit{Time of faster test} = \frac{\text{distance}}{\text{rate}} = \frac{0.19}{95} = 0.002 \text{ hour}$$

To convert to seconds, we use

$$\frac{0.002 \text{ hour}}{1} \times \frac{60 \text{ minutes}}{1 \text{ hour}} \times \frac{60 \text{ seconds}}{1 \text{ minute}} = 7.2 = 7 \text{ seconds}$$

Direct Variation

Another way in which problems make use of proportions is to refer to a *direct variation* among changing quantities; that is, when one aspect of a situation changes, another aspect changes in a proportional way or at the same rate.

Example 10

A construction firm knows that the number of new houses that it can build in a month varies directly with the cost of labor due to overtime charges. If 12 houses can be built when it pays an average of $14.50 per hour, how much should the company expect to pay its workers, if it needs to build 20 houses?

Solution 10

Since we are told that the quantities vary directly, we can immediately set up the model proportion.

$$\frac{\text{Number of houses}}{\text{Cost of labor}} = \frac{12}{4.50}$$

Let c be the cost of labor we are seeking. The model gives us

$$\frac{20}{c} = \frac{12}{14.50}$$

Cross-multiplying, we have $12c = 290$ and $c = 290/12 = \$24.17$ per hour.

Check: The check would be that the ratios would be the same, that is, $12 \div 14.50 = 0.83$ and $20 \div 24.17 = 0.83.\checkmark$ (Note that the calculator shows differences to the fourth decimal place. This is due to the rounding we did in the problem.) The ratio 0.83 is referred to as the *constant of variation*.

Example 11

The force applied to stretch an elastic spring varies directly with the amount that it is stretched. If a force of 15 dynes (the metric measure of force) stretches a spring 2.3 cm, how much force, to the nearest dyne, would be required to stretch the spring 6 cm?

Solution 11

Since we are told that the quantities vary directly, we can immediately set up the model proportion:

$$\frac{\text{Amount of force}}{\text{Stretch}} = \frac{15}{2.3}$$

Let f be the required force, and we have

$$\frac{f}{6} = \frac{15}{2.3}$$

and $f = 90/2.3 = 39.13 = 39$ dynes.

Check: The constant of variation must be the same: $15 \div 2.3 = 6.5$ and $39 \div 6 = 6.5.$ ✓

Percentage

Problems that involve percentages are really problems involving proportions. When asked for a percentage of a number, you can think of it as asking "How much would we have if we scaled our base amount to be 100?"

For example, when asked "What is 35% of 80?" we can translate this to mean "If we have 35 objects out of a set of 100, how much would we have in a *similar* group of 80?" The problem is then easily modeled by the proportion $35/100 = x/80$. This gives us $100x = 35 \times 80 = 2800$ and $x = 28$. Percentages appear in many of the problem situations we have seen before.

Investment

Example 12

When Gary invested $1200 in the Lion Fund, he had $1240 at the end of 3 months. Assuming the same rate of return, what would be the annual rate of return for this fund?

Solution 12

In order to determine the annual rate, we have to make use of the given assumption and conclude that over the year he would earn 4 times the interest received, since 3 months is one-fourth of a year. In other words, $4 \times \$40 = \$160$. Using the percentage model, we have

$$\frac{160}{1200} = \frac{x}{100}$$

Cross-multiplying gives us $1200x = 16,000$ and $x = 16,000/1200 = 13\frac{1}{3}\%$.

Check: We can calculate the interest for one year at $13.33\ldots$ percent and, then multiply by 0.25 to determine what the quarterly interest is; that is, $\$1200 \times 0.133333 \times .25 = \$39.9999$, which is approximately $40. (The difference arose because we did not have to use the entire repeating decimal.)$\checkmark$

Area

Example 13

The floor of a social hall that is 45 ft by 82 ft is carpeted except for a circular dance floor in the middle that has a diameter of 20 ft. To the nearest tenth of a percent, what percent of the floor is carpeted?

Solution 13

The solution requires finding the area of the entire floor, the area of the circle, and their difference, which is the amount of carpeted space. The percentage model is then used to give us

$$\frac{\text{Area of carpeted space}}{\text{Area of entire room}} = \frac{x}{100}$$

The area of the entire room is $45 \times 82 = 3690$ ft^2. The radius of the dance floor is 10 ft and using 3.14 for π, the area of the circular dance floor is $3.14 \times 10^2 = 314$ ft^2. Therefore, the carpeted area is $3690 - 314 = 3376$ ft^2.

The model gives us the proportion

$$\frac{3376}{3690} = \frac{x}{100}$$

Solving for x, we have $x = 3376 \times 100/3690 = 91.49 = 91.5\%$.

Check: The check is simple: $3690 \times 0.915 = 3376$ (rounded).✓

More complicated problems involving percentages can be found in *mixture problems* involving liquids that contain dissolved ingredients such as chemicals.

Example 14

A certain chemical solution used in manufacturing batteries contains water and acid. Initially, the solution is made with 50 kg of dry acid and 900 L of water. The solution is mixed and left to sit so that the water evaporates until it becomes a 30% acid solution. How much water, to the nearest liter, has to evaporate for this to be accomplished?

Solution 14

First, we need to make sure that our units of measurement are the same. Fortunately, in the metric system, 1 L of water weighs exactly 1 kg.

Regarding 30% as 30/100, the model for the problem is the proportion

$$\frac{\text{Amount of acid}}{\text{Remaining water} + \text{amount of acid}} = \frac{30}{100}$$

Let x be the amount of water that has evaporated. Therefore, $900 - x$ is the amount of remaining water and the total weight, in kilograms, is $900 - x + 50 = 950 - x$. The model gives us the equation:

$$\frac{50}{950 - x} = \frac{30}{100}$$

Cross-multiplying, we have $30(950 - x) = 5000$.

Simplifying and solving, we have

$$28{,}500 - 30x = 5000$$
$$23.500 = 30x$$
$$x = 23{,}500/30 = 783\frac{1}{3}$$

Therefore, 783 L of water has to evaporate.

Check: The remaining water is $900 - 783 = 117$, and the total weight is 50 kg of acid + 117 kg of water = 167 kg, that is, $50/167 = .299 = 29.9\%$, which is okay since we used a rounded amount.✓

Summary

1. When given a ratio of two or more objects, assign a variable to be the multiple of each number in the ratio and use the specific multiples given by the ratio as the amounts. Model the problem and substitute these variable amounts to derive an equation to solve. Solve for the multiple and compute the actual amounts.
2. Identify when a proportion is implied in the problem. This will be the case when phrases such as "in the same ratio as," "varies directly as," or "is proportional to" appear in the problem.
3. Shapes are similar only when corresponding sides are in proportion.
4. Computing a percentage or using a given percentage will usually be solved by creating a proportion in which the ratio of the actual amounts is equal to $x/100$, where x is the percentage.

Additional Problems

1. Separate 133 into two parts so that the ratio of the larger to the smaller is 4:3.
2. On a box for a model-airplane kit, labels indicated that the scale used is 1:72. If the wingspan of the model is 9 in, how long is the wingspan for the actual plane to the nearest foot?
3. CD Emporium finds that CDs of rock artists outsold CDs of classical music by 5:2 during November. If the profit from a rock CD is $2.50 and the profit from a classical CD is $1.80, how much profit was made if 2345 rock and classical CDs were sold during November?
4. A survey to determine attitudes of students of mathematics was given to 420 sixth-grade students 5 years ago and to the same group now, when they are in eleventh grade. In the sixth grade, the ratio of students who found math easy to those who didn't was 3:1; now, in the eleventh grade, it is 3:2. How many fewer students find math easy now than before?
5. Glenn High School has 1620 students in which the girl:boy ratio is 5:4. Of this group, 8% of the girls and 5% of the boys went on a field trip. What percent, to the nearest tenth, of the school's population went on this trip?

6. The ratio of the longer side of a rectangle to the shorter side is 7:2. The ratio of another rectangle's longer side to its shorter side is 4:3. Given that the longer side of the second rectangle is 11 in more than the longer side of the first and the shorter side is 18 in more than the shorter side of the first rectangle, find the ratio of the perimeters of the first rectangle to the second.

7. On a scale model of an outdoor patio at the home design center, there is an equilateral triangle in its center that is to be filled with a specially colored concrete mix costing $15.25 per bag, and each bag can be used to cover 8 ft². The model is scaled at a ratio of 1:8 compared to the actual size of the patio, and a side of the triangular region is 1.5 feet. How much will it cost to fill the region on the actual patio?

8. The sum of the digits of a two-digit number is 12. If the digits are reversed, a larger number is formed and the ratio of the larger number to the smaller number is 7:4. Find both numbers.

9. The ratio of the ages of a girl, her mother, and the girl's grandmother is 2:5:8. Six years ago the ratio of the girl's age to her mother's age was 4:13. How old are the three women now?

10. Carl's crew can complete 7 jobs in the same time that Ben's crew can complete 5 jobs. During the same week, Carl's crew completed 12 more jobs than Ben's crew. How many jobs were completed by both crews?

11. Fifty liters of an acid solution contains 18 L of pure acid. How many liters of acid must be added to make a 4% acid solution?

12. The amount of calories in a serving of ice cream varies directly with the amount of sugar the ice cream contains. A company makes a regular chocolate ice cream and a "Lite" version. Regular chocolate ice cream has 240 calories (cal) per serving and contains 64 g of sugar. How many grams of sugar does the Lite chocolate have, if it has only 180 cal per serving?

13. Mrs. Simmons tries to be a shrewd investor and invested her $28,000 in three different stocks in the ratio of 3:4:7. Over the course of a year, the respective returns were 8, 12, and 10%. What was her overall percentage return for the year?

14. A plane can make the same 1000-mi trip as an express train in 6 hours' less time. The ratio of their speeds is 5:2. At what speed does each vehicle travel?

Solutions to Additional Problems

1. Let x be the larger part of 133. Therefore, $133 - x$ is the smaller part. The problem gives us the exact proportion $x : 133 - x = 4 : 3$. Using the product of the means is equal to the product of the extremes, we have

$$4(133 - x) = 3x$$
$$532 - 4x = 3x$$
$$532 = 7x$$
$$x = 532/7 = 76$$

The larger part is 76, and the smaller part is $133 - 76 = 57$.

Check: $76 + 57 = 133$ and $76/57 = 1.333\ldots = 1\frac{1}{3} = \frac{4}{3}.\checkmark$

2. This is a simple problem that will use the following proportion as its model.

$$\frac{\text{Wingspan of model}}{\text{Actual wingspan}} = \frac{1}{72}$$

We can find the actual length in inches and then convert to feet. Let x be the number of inches in the actual wingspan. Therefore, we have $9 \text{ in}/x = \frac{1}{72}$. Cross-multiplying, we have $x = 9 \times 72 = 648$ in.

$$\frac{648 \text{ in}}{1} \times \frac{1 \text{ ft}}{12 \text{ in}} = 54 \text{ ft}$$

Check: We need to make sure that the ratio of the actual span to the model span is 1:72 or 0.0139. Using the calculated inches, we have $9 \div 648 = .0139.\checkmark$

3. We need to first find the amount of each type of CD sold. The total profit is the sum of the profits from each type. The model for the problem consists of two steps.

Step 1 Number of rock CDs + the number of classical CDs = 2,345.

Step 2 Total profit = the number of rock CDs × $2.50 + the number of classical CDs × $1.80.

Let x = multiple of 2 and 5

$2x$ = number of rock CDs sold in November

$5x$ = number of classical CDs sold in November

From the model for step 1

$$2x + 5x = 2345$$
$$7x = 2345$$
$$x = 2345 \div 7 = 335$$

Therefore, $2 \times 335 = 670$ classical CDs were sold and $5 \times 335 = 1675$ rock CDs were sold. From the model for step 2

$$\text{Total profit} = 1675 \times \$2.50 + 670 \times \$1.80$$
$$= \$4187.50 + \$1206.00 = \$5393.50$$

Check: The check is concerned mostly with the amounts of each type of CD sold. We need to make sure that the ratio is indeed 5:2 or 2.5, specifically, $1675 \div 670 = 2.5.\checkmark$ It is also necessary to recheck the arithmetic for step 2.

4. We have to compute the actual numbers of students who find math easy and those who do not at each grade level on the basis of the given ratios. For the sixth grade, let x be the multiple, $3x$ be the number of students who found math easy, and $1x$ be the number of students who did not.

$$3x + x = 420$$
$$4x = 420$$
$$x = 105$$

Therefore, at the sixth grade, there were $3 \times 105 = 315$ students who found math easy (and 105 who did not). At the eleventh grade (we should use a different variable so to avoid confusion between our two answers),

Let y = multiple

$3y$ = number of students who find math easy

$2y$ = number of students who do not

$$3y + 2y = 420$$
$$5y = 420$$
$$y = 84$$

Therefore, at the eleventh grade there are $3 \times 84 = 252$ students who find math easy (and $2 \times 84 = 168$ who do not). The difference $315 - 252 = 63$ is our answer.

Check: The check involves making sure that the numbers we determined are in the ratio given in the problem. The sixth-grade ratio should be 3 : 1 or 3, that is, $315 \div 105 = 3.\checkmark$ The eleventh-grade ratio should be 3:2 or 1.5, that is, $252 \div 168 = 1.5\checkmark$

5. The problem must be solved by finding the number of girls and boys who went on the trip, using the percentage model

$$\frac{\text{Number of girls on trip} + \text{number of boys on trip}}{\text{Total number of students}} = \frac{x}{100}$$

Realizing this, we see that simply adding $8\% + 5\% = 13\%$ is most likely *not* the correct answer. (This is the most common error in a problem like this.) It would be helpful to have an estimate before beginning. The ratio 5:4 indicates that slightly more than half of the students are girls. If the groups were equal, there would be 810 girls and 810 boys. Using 10% as an estimate for 8%, about 80 girls would be on the trip, and 5% of the boys would mean 40 boys on the trip. An estimate of the total number of students on the trip would be 120 and $120 \div 1620 = 0.074$, which is approximately 7 percent.

The algebraic scheme for finding the ratios will give us the numbers of girls and boys in the school.

$$\text{Let } x = \text{multiple}$$
$$5x = \text{number of girls}$$
$$4x = \text{number of boys}$$
$$5x + 4x = 1620$$
$$9x = 1620$$
$$x = 180$$

There are $5 \times 180 = 900$ girls and $4 \times 180 = 720$ boys. Thus 8% of the girls is $900 \times 0.08 = 72$ and 5% of the boys is $720 \times 0.05 = 36$. Therefore, $72 + 36 = 108$ students went on the trip. The model gives us the proportion: $108/1620 = x/100$ and $x = 10,800 \div 1620 = 6.66\ldots = 6.7\%$ rounded to the nearest tenth.

Check: The best check for a problem with these many steps is to recheck the arithmetic at each step and to feel sure that the answer is reasonable. The answer is reasonable on the basis of our initial estimate.

6. Since the only information about the first rectangle is the ratio of the sides, we will need to use the information about the second rectangle to help find the sides. Let x be the multiple needed for the first rectangle, $7x$ be the longer side of the first rectangle, and $2x$ be the shorter side. Then $7x + 11$ is the longer side of the second rectangle and $5x + 18$ is the shorter side of the second rectangle. We can set up the following proportion for the second rectangle.

$$\frac{7x + 11}{2x + 18} = \frac{4}{3}$$

Cross-multiplying, we have the equation $3(7x + 11) = 4(2x + 18)$. Simplifying and solving, we obtain

$$21x + 33 = 8x + 72$$

$$13x = 39$$

$$x = 3$$

Therefore, the dimensions of the first rectangle are $7 \times 3 = 21$ in and $2 \times 3 = 6$ in. Its perimeter is $2 \times (21 + 6) = 54$ in. The sides of the second rectangle are $21 + 11 = 32$ in and $6 + 18 = 24$ in. Its perimeter is $2 \times (32 + 24) = 112$ in. The ratio we want is 54:112 or 54/112, which reduces to 27/56 or 27:56.

Alternative Solution After having assigned the variable and obtained representations for the sides, we could have set up our answer in variable form immediately. Specifically, the perimeter of the first rectangle is $2(7x + 2x) = 18x$ and the perimeter of the second rectangle is $2(7x + 11 + 2x + 18) = 18x + 58$. The ratio of the perimeters would be $18x/(18x + 58)$. Once we found that $x = 3$, from the proportion, the ratio is quickly seen as 54/112.

7. The ratio plays only a small part in this problem. A scale of 1:8 simply means that the actual patio will be 8 times larger than the model. Therefore, the actual side of the triangle will be $8 \times 1.5 = 12$ ft. Using the formula for the area of an equilateral triangle from the reference table in Chap. 1, we find that the area is $\sqrt{3}/4 \times 12^2$ or approximately $1.73 \div 4 \times 144 = 62.28$ ft². Since each bag covers 8 ft², we will need $62.28 \div 8 = 7.785$ bags, which means that we have to buy 8 full bags. The cost is $15.25/bag \times 8$ bags = $122.00.

8. A simple way to solve the problem is to try all possibilities. The only two-digit numbers whose sum of digits is 12 and which are smaller than the number whose digits are reversed are 39, 48, and 57. The ratio 7:4 is to equivalent to 1.75. The ratios to consider are 93/39, 84/48, and 75/57. The only one of these equal to 1.75 is 84/48. Therefore, the numbers are 48 and 84.

An algebraic solution to problems involving the digits of a number are best solved by using two variables, one for the tens digit and one for the units digit. We know that the original number is the smaller of the numbers. Let u be the units digit and t be the tens digit of the original number. Therefore, $10t + u$ is the original number and $10u + t$ is the number when the digits are reversed. The problem gives us two pieces of information to use: (a) $t + u = 12$ and (b) $(10u + t)/(10t + u) = 7/4$. Using (b) and cross multiplying, we have $4(10u + t) = 7(10t + u)$. Simplifying gives us $40u + 4t = 70t + 7u$. From (a) we see that $t = 12 - u$ and when substituted into the equation, we have an equation in one variable to solve

$$40u + 4(12 - u) = 70(12 - u) + 7u$$
$$40u + 48 - 4u = 840 - 70u + 7u$$
$$36u + 48 = 840 - 63u$$
$$99u = 792$$
$$u = 792/99 = 8 \text{ (and } t \text{ must be } 12 - 8 = 4)$$

Therefore, the original number is 48 and the larger number is 84.

Check: We need to ensure that the ratio of the numbers is $7/4 = 1.75$, or $84/48 = 1.75.\checkmark$

9. An algebraic solution is the fastest way to the answer. Let x be the multiple for the current ages, $2x$ be the girl's age, $5x$ be the mother's age, and $8x$ be the grandmother's age. Six years ago, the girl's age was $2x - 6$ and the mother's age was $5x - 6$. With these, we have the proportion

$$\frac{2x - 6}{5x - 6} = \frac{4}{13}$$

Cross multiplying gives us the equation $13(2x - 6) = 4(5x - 6)$. Simplifying and solving, we have

$$26x - 78 = 20x - 24$$
$$6x = 54$$
$$x = 9$$

Currently, the girl is $2 \times 9 = 18$ years old, the mother is $5 \times 9 = 45$ years old, and the grandmother is $8 \times 9 = 72$ years old.

Check: Six years ago, the girl and her mother were 12 and 39 years old, respectively. The quotient ratio 12/39 is reducible to 4/13.✓

10. The first sentence of the problem gives us the ratio

$$\frac{\text{Jobs completed by Carl's crew}}{\text{Jobs completed by Ben's crew}} = \frac{7}{5}$$

Let x be the multiple, $7x$ be the number of jobs completed by Carl's crew, and $5x$ be the number of jobs completed by Ben's crew during the week. The second sentence gives us the simple model

Number of jobs completed by Carl's crew =
number of jobs completed by Ben's crew + 12

Inserting the variables, we have the equation $7x = 5x + 12$ and, clearly, $x = 6$. Therefore, Carl's crew completed $7 \times 6 = 42$ jobs and Ben's crew completed $5 \times 6 = 30$ jobs. The total number of jobs was 72.

Check: We need to ensure that 42:30 is equivalent to 7:5. We find that 42/30 does reduce to 7/5.✓

88

11. To better understand the problem, it is helpful to realize that the original solution is 36 percent acid; that is,

$$\frac{\text{Amount of acid}}{\text{Amount of acid} + \text{water}} = \frac{18}{50} = \frac{36}{100} = 36\%$$

Since we are adding more acid, we let x be the amount of acid added. This enables us to create the proportion $(18 + x)/(50 + x) = 40/100$. Cross-multiplying gives us the equation $100(18 + x) = 40(50 + x)$. Simplifying and solving for x, we obtain

$$1800 + 100x = 2000 + 40x$$
$$60x = 200$$
$$x = 3\tfrac{1}{3} \text{ L}$$

Check:

$$\frac{18 + 3\tfrac{1}{3}}{50 + 3\tfrac{1}{3}} = \frac{21\tfrac{1}{3}}{53\tfrac{1}{3}} = \frac{64}{3} \div \frac{160}{3} = \frac{64}{3} \times \frac{3}{160}$$
$$= \frac{64}{160} = 0.4 = 40\% \checkmark$$

12. The phrase "varies directly" indicates that the ratios of calories to amount of sugar are proportional. Therefore, letting g be the number of grams of sugar in the Lite version, we have the proportion $240/64 = 180/g$. Cross multiplying gives us $240g = 180 \times 64 = 11,520$ and $g = 11,520 \div 240 = 48$ g of sugar.

Check: $240/64 = 3.75$ and $180/48 = 3.75. \checkmark$

13. Note the word *respective*. This is usually meant as a direction to make a correspondence between the first items in each list, the second items in each list, and so on. In this problem, it means that the smallest investment earned 8%, the middle amount earned 12%, and the largest investment earned 10%. Each percentage return must be calculated for the money in that investment. Therefore, we have to solve this problem in several steps.

(a) Find the amount of each investment. Let x be the multiple. The three amounts invested are $3x$, $4x$, and $7x$. Therefore, $3x + 4x + 7x = 14x = 28,000$ and $x = 2000$. The three amounts are actually $6000, $8000, and $14,000.

(b) Find the new amounts on the basis of the percentage returns. We can use the formula $A = p + prt$, where $t = 1$. For the $6000 investment, she earned $6000 \times .08 = \$480$. For the $8000 investment, she earned $8000 \times .12 = \$960$. For the $14,000 investment, she earned $14,000 \times .10 = \$1400$. The total return was $\$480 + \$960 + \$1400 = \$2840$.

(c) Find the overall percentage using $2840/28,000 = x/100$, that is, $x = 2840 \times 100 \div 28,000 = 10.14\%$.

Check: The most important aspect of the problem to check is the amounts of each investment represented in the ratio of 3:4:7. Checking for a ratio involving three or more terms can be done pairwise: $\$8000/\$6000 = 4/3$, $\$14,000/\$8000 = 7/4$. Both are true. The rest of the check would be to recheck the arithmetic in finding the earnings and overall percentage.✓

14. We can use two variables to set up this problem quickly. Let x be the multiple for the speeds. Since the plane is obviously the faster vehicle, $5x$ is the speed of the plane and $2x$ is the speed of the train. Let t be the time of the train and, therefore, $t - 6$ is the time of the plane. Using Distance = rate × time, we have $2xt = 1000$ or $xt = 500$, and realizing that the distances traveled by each are equal, we have the equation $2xt = 5x(t - 6)$. Simplifying gives us $2xt = 5xt - 30x$ or $30x = 3xt$. Using $xt = 500$ from above, we have $30x = 1500$ and $x = 50$.

Therefore, the train travels at $2 \times 50 = 100$ mph and the plane travels at $5 \times 50 = 250$ mph.

Check: We need to check the ratios of the speeds and the time taken by each vehicle. The quotient ratio $250/100$ can be reduced to $5/2$.✓ Using Time = distance ÷ rate, the time taken by the train is 1000 mi $\div 100$ mph $= 10$ hours, and the time taken by the plane is 1000 mi $\div 250$ mph $= 4$ hours, which is 6 hours less.✓

must always make sure to *examine the proper statistics*. Here, it might be much more helpful to simply look at the statistical records of the games the teams played, against each other and against all other opponents. And then, of course, there is always the matter of things like talent, athletic ability, strategy, coaching, teamwork, and just plain common sense on the court!

Make use of statistics; rely on statistics. But make sure you are using the *proper statistic*, at the *proper time*, and in the *proper context*.

MODEL PROBLEM 2

Laura has received test grades of 92, 97, 91, 90, and 88 this marking period. What is her average? What grade would she need on her next test to have an average of 95?

Laura's average at this time is simply $\frac{92+97+91+90+88}{5}$ or $\frac{458}{5}$ or **91.6.**

Since we find the average by adding all her scores and then dividing by the number of scores, for her to have an average of 95, on what will be 6 tests, she would need a total of 6×95 points. That means that she would need 570 points. She now has a total of 458. How many more points does she need? She would need, on her next test, $570 - 458$, or **112.** Will she be able to reach her goal of a 95 average with just one more test? If, as usual, the maximum test score she could achieve is 100, *she will not be able to do it with just one more test.*

What would be the highest her average could be after one more test?

Since the most she could receive on the next test is 100, her new average would be $\frac{458+100}{6}$ or $\frac{558}{6} = $ **93.** (Remember, there is no need to do all the addition again, since we already know the total of the first five test scores.)

MODEL PROBLEM 3

Helen's mother promised her a reward if she can just bring up her math scores so that she is in the top half of her class. On the last test, the test scores in her class were 97, 83, 94, 100, 95, 98, 93, 75, 65, 93, 97, 83, 85, 95, 82, 100, 84, 95, 99,

94, 99, 92, 94, 95, 100, 82, 84, 65, 88, 74. Helen scored a 95. Did she make her goal? Is she due her reward?

First, let's decide which of the measures of central tendency will help us to solve this problem. Since we are really interested in Helen's *standing*, or *place*, in the class, the *median* will be of the most help to us. For that, we must first put the scores in (numerical) order.

In numerical order, we have:

100	98	95	93	84	82
100	97	95	93	84	75
100	97	94	92	83	74
99	95	94	88	83	65
98	95	94	85	82	65

Since we have 30 scores, the median would be the *average* of the two *middle scores*. The two middle scores would be the fifteenth and sixteenth scores on the list. We can spot them fairly easily with the scores arranged the way they are above. The fifteenth and sixteenth scores are 94 and 93. The average of 94 and 93 is 93.5. Therefore, **the median is 93.5.** Helen's score is *above* 93.5, and therefore she is in the top half of her class and is entitled to her reward. [We could actually SEE that she is in the top half if we examine the arrangement above, since the 95 is in one of the first three (of six) columns.]

Practice Problems Involving Statistics

1. Harolyn has received test scores of 97, 99, 92, and 91. What score must she receive on her next test to have an average of 95?
2. Tommy has received test scores of 93, 100, 88, and 74. Can he achieve an average of 90 with his next test score? If so, what grade would he have to receive?
3. Sharon was promised a bracelet if she can achieve an average of 95 in math this quarter. So far, her test grades are 93, 95, 97, 88, and 95. She has one more test this quarter. Can she get her bracelet? Explain why or why not.
4. Beach Middle School gives a special award to basketball players who can *average* 30 points per game AND maintain a 90 average in their major subjects. Given the following information, which (if any) of these students would be eligible for that special award?

NAME	POINTS/GAME	GRADES/MAJOR SUBJECTS
Joey	30, 23, 37, 31, 29	85, 88, 97, 97, 83
Jimmy	60, 20, 10, 33, 29	65, 99, 99, 89, 99
Jerry	60, 30, 20, 30, 0	90, 89, 91, 92, 88
Justin	29, 31, 30, 30, 30	99, 66, 80, 95, 98
Jorge	45, 45, 10, 15, 45	90, 90, 93, 85, 93

5. Nadine and Ruben always seem to be in competition with their grades. So far this year, their grades have been as follows:

   Nadine: 97, 65, 92, 100, 85, 94, 93, 67, 70, and 87
   Ruben: 85, 86, 84, 86, 84, 85, 87, 83, 85, and 85

   What are their respective averages? If you had to "bet" on one of them keeping that average, which one would be the likelier candidate? Why?

6. Which of the following students will be eligible for induction into the Honor Society? The requirements for the Honor Society are an average of 88 AND an average of 35 hours of community service, over the last 4 years.

NAME	AVERAGE EACH YEAR	NUMBER OF HOURS OF COMMUNITY SERVICE
Jennifer	87, 89, 91, 88	37, 40, 23, 46
Jeanne	92, 90, 85, 88	27, 37, 36, 39
Jane	85, 83, 90, 88	54, 57, 45, 40
Jessica	87, 89, 88, 82	40, 32, 34, 30

7. In the school mentioned above, there is also a Credit Society. The requirements for the Credit Society are an average of 88 OR an average of 35 hours of community service, over the last 4 years. Which of the girls mentioned in the last problem would be eligible for the Credit Society?

8. Charlie is on a hockey team outside of school. He would like to be selected to be on the county team. For this, he needs a goals per game average of 2.5 and an assist per game average of 1.5. So far this year, he has the following record:

   Goals: 1, 0, 2, 0, 4, 2, 3, 3, 3
   Assists: 1, 0, 1, 0, 2, 1, 1, 2, 3

   How many goals and how many assists does he need in the last game to qualify for the county team?

9. Find the mean, median, mode, and range for the following set of scores:

$$88, 92, 88, 91, 88, 97, 98, 90, 90$$

10. Figure out **your** average this year. Is it as high as you would like? Decide on a **goal** for yourself. What grade would you need on your next test to reach it? Can you reach it on the next test, or might you need another? Is it attainable this year?

Solutions to the Practice Problems Involving Statistics

1. If Harolyn is to achieve the 95 average she wants on the next test, it would mean she would have to have a total of (95)(5) points, or 475 points. Her grades so far are 97, 99, 92, and 91, for a total of 379 points. Therefore, for her 95 she would need 475 − 379, or **96 points on her next test.**

2. Tommy wants an average of 90 after his fifth test, so he would need a total of (90)(5) points, or 450 points. His grades so far are 93, 100, 88, and 74, for a total of 355 points. Therefore, he would need 450 − 355, or **95 points on his next test.**

3. Sharon would like an average of 95 after her sixth test, so she would need (95)(6) points, or 570 points. With her 93, 95, 97, 88, and 95, she has so far accumulated a total of 468 points. To get her 95 average (and her bracelet) she would therefore need 570 − 468, or **102 points.** It looks as though she will have to wait at least one more test for her bracelet!

4. Let's take a look, player by player.

NAME	AVERAGE OF POINTS	ACADEMIC AVERAGE
Joey	$\dfrac{30 + 23 + 37 + 31 + 29}{5} = 30$	$\dfrac{85 + 88 + 97 + 97 + 83}{5} = 90$
Jimmy	$\dfrac{60 + 20 + 10 + 33 + 29}{5} = 30.4$	$\dfrac{65 + 99 + 99 + 89 + 99}{5} = 90.2$
Jerry	$\dfrac{60 + 30 + 20 + 30 + 0}{5} = 28$	$\dfrac{90 + 89 + 91 + 92 + 88}{5} = 90$
Justin	$\dfrac{29 + 31 + 30 + 30 + 30}{5} = 30$	$\dfrac{99 + 66 + 80 + 95 + 98}{5} = 87.6$
Jorge	$\dfrac{45 + 45 + 10 + 15 + 45}{5} = 32$	$\dfrac{90 + 90 + 93 + 85 + 93}{5} = 90.2$

Since the boys need to qualify in BOTH areas, with a game point average of 30 and a 90 average in their majors, **Joey, Jimmy, and Jorge will qualify; Jerry and Justin will not.**

5. Nadine's average is $\frac{97+65+92+100+85+94+93+67+70+87}{10} = \frac{850}{10} = 85$

   Ruben's average is $\frac{85+86+84+86+84+85+87+83+85+85}{10} = \frac{850}{10} = 85$

   They both have the same average, but if *I* had to "bet" on one of them keeping that average, I'd have to go with Ruben, since his scores are all right around 85, while Nadine's have been as many as 18 points away. In other words, Nadine's *range* is $100 - 67$, or *33 points*, while Ruben's is $87 - 83$, or *4 points*.

6. Again, let's do this one student at a time.

NAME	ACADEMIC AVERAGE	AVERAGE NUMBER OF HOURS OF COMMUNITY SERVICE
Jennifer	$\frac{87 + 89 + 91 + 88}{4} = 88.75$	$\frac{37 + 40 + 23 + 46}{4} = 36.5$
Jeanne	$\frac{92 + 90 + 85 + 88}{4} = 88.75$	$\frac{27 + 37 + 36 + 39}{4} = 34.75$
Jane	$\frac{85 + 83 + 90 + 88}{4} = 86.5$	$\frac{54 + 57 + 45 + 40}{4} = 49$
Jessica	$\frac{87 + 89 + 88 + 82}{4} = 86.5$	$\frac{40 + 32 + 34 + 30}{4} = 34$

   Of the four girls, only **Jennifer** has met both requirements, and, therefore, **only Jennifer is eligible for the Honor Society.**

7. This time, **three** of the girls are **eligible.** Do you see the difference in the requirements? For the Credit Society a student is expected to meet the academic requirement **OR** the service requirement, and therefore is eligible if he/she meets **one (or both)** requirement(s).

8. In the 9 games played so far this year, Charlie has scored $1 + 0 + 2 + 0 + 4 + 2 + 3 + 3 + 3$, or 18 goals. To achieve an average of 2.5 goals this year, Charlie would need a total of $(2.5)(10)$, or 25 goals. That means he would need $25 - 18$, **7 goals.** (Do you think he can do it?)

   Let's see if he's any closer with the assists. For an assist per game average of 1.5, he would need $(1.5)(10)$, or 15 assists. In the first 9 games, he has a total of $1 + 0 + 1 + 0 + 2 + 1 + 1 + 2 + 3$, or 11 assists. Therefore, he would need "only" $15 - 11$, or **4 assists.** (That may not seem like a lot, but notice he has never scored that many in one game before. But maybe his desire to get on the county team will inspire him to really great play in his last game!)

9. The scores were 88, 92, 88, 91, 88, 97, 98, 90, and 90.

The **mean** would be $\frac{88+92+88+91+88+97+98+90+90}{9} = \frac{822}{9} = 91.3$.

To find the **median**, we will first have to arrange the scores in (numerical) order:

| 88 | 88 | 88 | 90 | **90** | 91 | 92 | 97 | 98 |

The **median**, or middle, score **is 90.**

The **mode** would be **88,** since there are more 88's than any other score.

The **range** would be the difference between the highest score and the lowest score, or

$$98 - 88, \text{ or } 10 \text{ points.}$$

10. I can't help you with this one, but **I wish you the best of luck in achieving your goal.**

Chapter 9

Number Problems

Sometimes the numbers we deal with don't have "stories" behind them. They are "just numbers," numbers that are related to one another in special ways. There are many ways to do these problems. Usually, they can be solved algebraically, arithmetically, and/or logically. Let's start with some *very simple* problems that you can almost do in your head. That should help us just to see the proper procedures and the various options open to us.

MODEL PROBLEM I

I am thinking of two numbers. One number is 7 more than the other. The sum of the numbers is 29. Find the numbers.

"Guess and check" works very nicely with something like this. Select any number, but make sure it is a *logical* choice. If it is one of two numbers whose sum is 29, it must be less than 29. If it is to be the smaller of the two numbers, it must be less than half of the 29. Therefore, pick a number less than half of 29. What number is 7 more than your number? Now add the two numbers together. Was the sum 29? Were you lucky enough to have guessed the correct number to use as a starting point?

For example, let's try a nice round number, such as 10. Seven more than 10 is 17. If we add 10 and 17, we get 27. This is *not quite* what we want. So, the next number we would try is 11. Seven more than 11 is 18. If we add 11 and 18, we get 29. This *is* the sum we need. Therefore, **the required numbers are 11 and 18.** We could have done the problems algebraically as well. If we let **x** represent the smaller number, then the larger one must be **x + 7.**

$$x + x + 7 = 29$$

Combining like terms: $\quad 2x + 7 = 29$

Subtracting 7: $\qquad\qquad \underline{-7 \quad -7}$

$$2x = 22$$

Dividing by 2: $\qquad\qquad \dfrac{2x}{2} = \dfrac{22}{2}$

$$\mathbf{x = 11 \ (smaller\ number)}$$
$$\mathbf{\underline{x + 7 = 18} \ (larger\ number)}$$

Check: $\qquad\qquad\qquad\qquad 29$

MODEL PROBLEM 2

I am thinking of two numbers. The sum of the numbers is 57. Their difference is 13. What are the numbers?

Let's look at several methods of solving this one. Again we'll start with "guess and check." Again, however, let's use *logic* and make *reasonable* guesses. Since the sum must be 57, the numbers must be less than 57. Since the difference is 13, the numbers are *not* right around half of 57. Since half of 57 is approximately 28, and we want less than that, try a number a little less than 28.

Suppose we again try a "round number," like 20. If the sum of the two numbers has to be 57, the "other number" has to be 37. The difference between 37 and 20 is 17. That's too much. So, let's try 21. The "other number" now would have to be 36. The difference now is 15. We're getting closer. (And if we look at the differences we have found so far, we see that increasing from 20 to 21, the difference went down by 2. Another increase of 1 in our guess should bring down the difference another 2. That would be just right!) Trying 22, we find that the "other number" would have to be 35. The difference between 35 and 22 *is* 13, so **our numbers must be 22 and 35.**

Actually, we could have made a *very educated* initial guess by looking at the second condition as well. Since the difference between the numbers is 13, we can try "splitting" that, and try 6 or 7 less than 28 as our initial guess. This method would tell us to try 22 or 21 right from the start. (In this particular problem, it wouldn't have saved us a whole lot of time, but if we had had to try several "guesses" before finding the one that "worked," it might have made a big difference to take that second condition into account as well.)

This problem could have been done algebraically as well, using either one or two variables. Suppose we let X represent one of the numbers. If the difference between the numbers is 13, it means that one number is 13 more than the other. In that case, it is easier to let X represent the *smaller* number. We can then represent the larger number by $X + 13$.

Since the sum of the numbers is 57:	$X + X + 13 = 57$
Combining like terms:	$2X + 13 = 57$
Subtract 13:	$\underline{-13 \quad -13}$
Divide by 2:	$\dfrac{2X}{2} = \dfrac{44}{2}$
	$\mathbf{X = 22}$
	(smaller number)
AND	$\mathbf{X + 13 = 35}$
	(greater number)

I think that, probably, you have not worked with two equations in two unknowns. However, in this case, doing the problem that way can make the solution of the problem very simple. So, I'll show you that method. If you don't like it, don't feel up to it, or don't follow it, don't worry. Use either of the other methods.

Let X = greater number
Y = smaller number

Since the sum is 57:	$X + Y = 57$
Since the difference is 13:	$\underline{X - Y = 13}$
Add the equations:	$2X = 70$ (Notice that since we added Y and $-Y$, the Y's "dropped out," and we now have only one variable.)
Divide by 2:	$\mathbf{X = 35}$ **(greater number)**

Substituting for X, we find that $\mathbf{Y = 22}$
(smaller number)

MODEL PROBLEM 3

Dennis was buying his school supplies and paid a total of $6.50 for a notebook and a pen. The pen cost $1 more than the notebook. What was the cost of each item?

We again have the option of an arithmetic or algebraic solution. There is a tendency, sometimes, with a problem like this, to look at the problem quickly, look at the numbers involved, and *immediately* give an answer. Unfortunately, I have found that most students try to do the problem too quickly, look at only one condition, and jump to the conclusion that that solution will satisfy BOTH conclusions, when, in fact, it satisfies one condition but not the other.

Let's "suppose" that the notebook cost $2; the pen would then cost $3 ($1 more). The total cost of the items would then be $5. That's too little. If we try $3 for the notebook, the pen would then be $4. That would make the total cost $7. That's too much. Suppose we try $2.50 for the notebook; the pen is now $3.50. The new total is $6. Closer, but still not correct. Since we are now only 50 cents under the correct total, we see that we would have to add less than 50 cents to our "guesstimate." Suppose we cut that in half and try 25 cents more. The notebook is then $2.75 and the pen $3.75. The new total: $6.50. We now know that **the cost of the notebook was $2.75, and the cost of the pen was $3.75.** Make a *reasonable* "guess," and then kind of jockey back and forth (too high, too low) and "close in" on your answer.

Algebraically,

$$\text{Let } X = \text{the cost of the notebook}$$
$$X + \$1 = \text{the cost of the pen}$$
$$X + X + \$1 = \$6.50$$
$$2x + 1 = 6.50$$
$$\underline{-1 \quad -1}$$
$$\frac{2X}{2} = \frac{5.50}{2}$$
$$X = 2.75$$
$$X + 1 = 3.75$$

The cost of the notebook is $2.75 and the cost of the pen is $3.75.

MODEL PROBLEM 4

What is the *smallest* number such that when it is divided by 2 or by 3 or by 4 or by 5, there is always a remainder of 1?

This is a problem where you certainly can use "guess and check," but a little logic and a little number theory will help a lot more.

What do we know about the number? We know it has to be odd. (If you aren't sure why, think about it, and read on!) We also know that when it is divided by 5, there is a remainder. Numbers divisible by 5 all end in 0 or 5. So, if when divided by 5, there is a remainder of 1, the units' digit would have to be 1 (1 more than a number ending in 0) or 6 (1 more than a number ending in 5). But the number MUST be odd (so that when divided by 2, there is a remainder of 1). So, we know that the units' digit of our number must be 1. We could, then, try 11, or 21, or 31, etc., but if we realize that all the numbers listed above must be *factors* of 1 less than our number, we should look for a *common multiple*. Since we want the *smallest* such number, we should look for the *least common multiple*.

[For those who need a refresher on finding the least common multiple, I'll remind you. (Those who don't need the refresher, just go right on to the next paragraph!) As the name suggests, *common multiples* are numbers which are multiples of the numbers we are considering. The *least* common multiple is the smallest of these. How do we find the least common multiple? We *could* list all the multiples of our numbers until we see one in both lists. The problem is that although we *might* find the "LCM" (least common multiple) quickly, we also *might not*. An easier way is to use the prime factors of our numbers. Then, all we would have to do is to multiply the prime factors of our numbers, making sure not to include factors already included. For example: suppose we are asked to find the LCM of 16 and 18. If we use the "list" method:

Multiples of 16: 16, 32, 48, 64, 80, 96, 112, 128, 144, 160, . . .
Multiples of 18: 18, 36, 54, 72, 90, 108, 126, **144,** . . .

We can stop now that we have found a number in both lists.

Using the prime factor method:

$$16 = 2 \times 2 \times 2 \times 2$$
$$18 = 2 \times 3 \times 3$$

To find the LCM, list the factors of one of the numbers, for example, $2 \times 2 \times 2 \times 2$. Now look at the factors of the other number. We need a 2. Do we have a 2? Since we do, go on to the next factor. We need two factors of 3. Since we have none, we must now include those, and our LCM becomes $2 \times 2 \times 2 \times 2 \times 3 \times 3$, or **144**.]

Getting back to our original problem (and for those of you who have skipped down to this point), *our least common multiple, then, can be found by multiplying the prime factors of our numbers, making sure not to include factors already included.* Our job is a little simpler if we realize that 2, 3, and 5 are all prime numbers. Only 4 is not. Its prime factors are 2 and 2. Since one of those two factors of 2 is already accounted for, the LCM is $2 \times 3 \times 2 \times 5$ or 60. Since we need a remainder of 1, **the number we are looking for is 61.**

Practice Number Problems

1. I am thinking of two numbers. The sum of the numbers is 81. Their difference is 9. Find the numbers.
2. The greater of two numbers is 25 more than the smaller. Their sum is 65. Find the numbers.
3. Find the smallest number such that when it is divided by 3 or 4 or 5 or 6, the remainder is 1.
4. The students at Baytown Middle School made jewelry boxes in wood shop. They made a total of 50 jewelry boxes. Some of the boxes had 3 drawers and some had 2 drawers. When the students finished sanding the drawers, they found they had a total of 118 drawers. How many of each type of jewelry box had they made?
5. Two numbers are in the ratio of $4:7$. Their sum is 176. Find the numbers.
6. Three numbers are in the ratio of $7:5:4$. Their sum is 384. Find the numbers.
7. The greater of two numbers is 5 times the smaller. Their sum is 276. Find the numbers.
8. Which would you prefer: a 10% raise at the end of <u>each</u> of 2 years, or a 20% raise at the end of the second year?

9. In Princetown Middle School, there are 5 girls for every 4 boys. If there are 1206 students in the school, how many boys are there and how many girls?

10. If there are twelve 1-cent stamps in a dozen, how many 4-cent stamps are there in a dozen?

11. This is an old problem, seen most recently on a TV show, where it was posed to a detective who often has to use logic to solve his cases. A man has 12 sacks of gold. Each sack of gold contains 18 coins. In one sack, however, the coins are counterfeit. Real coins weigh 1 pound each; counterfeit coins weigh 1 pound, 1 ounce. Liz wants to figure out which sack has the counterfeit coins, but she has a penny scale, and *only one* penny. How can she figure out which sack has the counterfeit coins?

Solutions to the Practice Number Problems

1. *Arithmetically:* Our numbers have to be *fairly* close, but not VERY close. Half of 81 would be about 40. So, we can try, say, 35. If the sum is to be 81, the other number would have to be 46. The difference between 46 and 35, though, is *11*. Not close enough, but *almost* what we want. So, let's try to get a *little* closer. Let's try 36. If the sum is to be 81, the other number would have to be 45. The difference between 45 and 36 is 9. This is just what we want. So, **the required numbers are 36 and 45.**

Algebraically:
One number must be 9 more than the other. So,

$$\text{Let } X = \text{smaller number}$$
$$X + 9 = \text{greater number}$$

The sum of the numbers is 81.	$X + X + 9 = 81$
Combine like terms.	$2X + 9 = 81$
Subtract 9.	$\underline{ -9 \quad -9}$
	$2X = 72$
Divide by 2.	$\dfrac{2X}{2} = \dfrac{72}{2}$
	$\mathbf{X = 36 \text{ (smaller number)}}$
	$\mathbf{X + 9 = 45 \text{ (greater number)}}$

2. *Arithmetically:* Half of 65 is about 32. Since the numbers are 25 apart, "split the difference" and try "about 12" less than 32. That makes our "guesstimate" 20. Since the other number must be 25 more, that would make the other number 45. Now add to see how close we are with the sum. Got it on the first try! *Since 20 and 45 differ by 25, and the sum of 20 and 45 is 65*, **the numbers we want are 20 and 45.**

Algebraically:

Let X = smaller number

Since the other
number is 25 more: $\quad X + 25 = $ greater number

Since the sum of
the numbers is 65: $\quad X + X + 25 = 65$

Combine like terms: $\quad 2X + 25 = 65$

Subtract 25 from
both members
of the equation: $\qquad\qquad -25 \quad -25$

Divide both members
of the equation by 2: $\qquad \dfrac{2X}{2} = \dfrac{40}{2}$

$$X = 20 \text{ (smaller number)}$$
$$X + 25 = 45 \text{ (greater number)}$$

(Check: $\qquad\qquad\qquad\quad 65 \qquad\qquad\qquad$)

3. The easiest way to do this problem is to find the LCM (least common multiple) of our numbers. If you consider each in terms of its prime factors, you will be able to find this easily. 3 and 5 are prime numbers. $4 = 2 \times 2$, and $6 = 2 \times 3$. The product of the prime factors, repeating only if *necessary* would then be $3 \times 5 \times 2 \times 2$ [we have no other factor(s) of 2.] Notice that both factors of 6 already appear. Therefore, we do not need any other factors to find the LCM. Multiplying, we find that $3 \times 5 \times 2 \times 2 = 60$. Since we need a remainder of 1, add 1 to our LCM. Therefore, the **smallest number that has a remainder of 1 when divided by 3, 4, 5, or 6** is **61.**

   [An interesting variation on this problem would be: Find the smallest number such that when it is divided by 3, the remainder is 2; when divided by 4, the remainder is 3; when divided by 5, the remainder is 4; when divided by 6, the remainder is 5. Notice that in each case the remainder is one less than the divisor. That means we are one "short" of having a factor. The LCM is the same (60), but in this case, we have to *subtract* 1, and, therefore, **the number we want is 59.**]

4. The easiest place to start might be 25 of each type of jewelry box. Let's arrange our findings in a table:

NUMBER OF 3 DRAWER BOXES	NUMBER OF DRAWERS	NUMBER OF 2 DRAWER BOXES	NUMBER OF DRAWERS	TOTAL
25	**75**	25	**50**	**125**

We aren't too far off. Since we want *fewer* drawers, it probably would not help us to add more boxes with the greater number of drawers. So let's try 24 and 26.

24	**72**	26	**52**	**124**

That (decreasing the number of 3-drawer boxes by 1) brought our total down, but only by 1. Since we need a total 6 less than our current one, it might follow to try decreasing the number of 3-drawer boxes by 6. That means that we should try 18 (and 32).

18	**54**	32	**64**	**<u>118</u>**

Therefore, the students had made **18 jewelry boxes with 3 drawers** and **32 jewelry boxes with 2 drawers**.

5. Once again, we have the option of doing the problem either arithmetically or algebraically (with a little logic thrown in).

Pick two numbers in the ratio of 4 : 7, say, 44 and 77. Their sum is 121. That sum is too small. We could try doubling, but if we did, the sum would be 242. That would give us way too much, so we know our numbers cannot be 88 and 154. We can keep working our way up from 44 and 77, but you might not want to bother with the ratios. So, what you can do is to take *another* pair in the ratio of 4 : 7, such as 40 and 70. That sum is 110, still too much if we double. Maybe we should try half of each number, 20 and 35. That sum is 55, and you might recognize that if we triple this, we get 165, *just a little under the desired sum*. Therefore, 60 and 105 (triple each of our numbers) would get us *very* close. In fact, we are only *11* under our desired sum. This represents one group of 4 and one of 7. Therefore, it would seem our numbers are **64** (add 4 to the 60) and **112** (add 7 to the 105). Their sum IS 176, as required, but you might want to just double check that the numbers are in the ratio 4 : 7.

Another arithmetic way of doing this problem is to reason that $4 + 7$ is 11. Each time we add another 4 and another 7, we get another 11. How many groups of 11 are there in 176? Since there are 16, there have to be *16 groups of 4* and *16 groups of 7*, or, in other words, **64** and **112.** This concept will help us with an algebraic solution.

There are longer ways, of course, with more trial and error and less logic and reasoning.

But let's look at an algebraic solution. If you remember, in Chapter 5, I mentioned something called the "common ratio factor" (commonly referred to as "CRF"). [Again, let's do a little refresher course for those who feel they need one (and, again, if you don't, just skip right down to the solution). A ratio, remember, is a comparison of two numbers. Sometimes, these two numbers have a common factor. When we express a ratio in simplest terms, in fact, we have divided both numbers by that common factor. (Consider the ratio of $32:48$, for example. Since 32 and 48 are both divisible by 16, we can divide both terms of the ratio by 16 and find that, in simplest terms, $32:48 = 2:3$, and 16 is the common ratio factor.) Looking at this problem from the "other direction," if we knew that 2 numbers were in the ratio of $2:3$, that would not be enough to tell us what the numbers are; there are an infinite number of pairs in the ratio of 2 to 3. BUT, what we DO know is that one of the numbers is 2 times *something* and the other number must be 3 times the *same thing*. That "*same thing*" is the "*common ratio factor.*"]

So, if we	Let $X = $ CRF
then	$4X = $ first number
and	$7X = $ second number
So,	$4X + 7X = 176$
Combining:	$\dfrac{11X}{11} = \dfrac{176}{11}$
Divide by 11:	
	$X = 16$

That's our common ratio factor, just like our "groups" of 11 in the solution above.

$$4X = 64$$

$$7X = 112$$

6. Decide which of the above methods you like best, and then use it for this very similar problem. Trial and error can get a little "messy" with three numbers, but it still is effective. However, it may be effective, but it does get very repetitious if you don't "guess" right. You'd be much better off using the "how many 'bundles' " concept, either arithmetically or algebraically.

The numbers are in the ratio $7:5:4$. Since $7 + 5 + 4 = 16$, we have 16 in a "group" or we have that multiple of the "common ratio factor."

Dividing 384 by 16, we get 24. (Our "common ratio factor" is 24.) Therefore, we want 24×7, 24×5, and 24×4.

Our numbers, then, are 168, 120, and 96. Adding, just to make sure, we get a total of 384.

7. "Guess and check," or solve this algebraically.

$$\text{Let } X = \text{smaller number}$$
$$5X = \text{greater number}$$
$$X + 5X = 276$$
$$\frac{6X}{6} = \frac{276}{6}$$
$$X = \mathbf{46}$$
$$\underline{5X = \mathbf{230}}$$
$$276$$

8. The simplest way is just to figure out where you would stand after each of these situations. Since we are asked simply *which* would be better, but not *how much* is involved, select any number to be your base. A nice, easy, safe number, when you are working with percents, is 100.

Let's see what happens in each situation.

Plan 1: A 10% raise after 1 year means a $10 raise after the one year. Under Plan 1, then, you are then making $110. Another 10% raise means an $11 raise. You are now making, at the end of 2 years, $121.

Plan 2: A 20% raise at the end of the second year means that you have a $20 raise, giving you $120.

Therefore, under Plan 1, you would be earning $121 at the beginning of year 3. Under Plan 2, you would be earning $120 at the beginning of year 3. Which would you prefer?

But let's also consider how much you would have earned to that point.

Under Plan 1, you earn $100 the first year and $110 the second year, giving you a total of $210 earned in the first 2 years. In year 3, you'll earn $121. This gives you a total of $331 after 3 years.

Under Plan 2, you earn $100 the first AND second years, for a total of $200 earned in the first 2 years. In year 3, you'll earn $120.

This gives you a total of $320. No matter how you look at it, "slow and steady wins the race" again; Plan 1 is better for you.

9. Try "groups of 5" and "groups of 4," if you like. Try 50 and 40, or 500 and 400. The latter, of course, is MUCH closer. It's closer, but it's still just 900. You *still* need more than 300. Well, every group of 50 and 40 gives you 90. Since we need over 300, let's take at least 3 "groups" of 50 and 40. That means that we should add 150 and 120, respectively, to the 500 and 400. That now gives us 650 and 520. That gives us a total of 1190. We are now only 36 short. Well, each "group" of 5 and 4 accounts for 9 students. We need 4 of these groups of 9, so we need 4 times our 5 and 4, or 20 and 16. Our 650 and 520, then, become **670** and **536.** Checking, $670 + 536 = 1206$.

10. I hope you READ this one carefully. Did you do any computation? I hope not! There are, of course, **12** of *anything* in a **dozen!!**

11. Since you can only weigh the coins once, there has to be a way of differentiating among the sacks right from the beginning. Suppose you number the sacks. Now take *that* number of coins from the sack. (For example, 1 coin from sack 1, 2 coins from sack 2, 3 coins from sack 3, etc.) The real coins will all weigh 1 pound. But the counterfeit coins will weigh 1 pound, 1 ounce. First of all, how many coins are you weighing? You took **1** coin from the first sack, **2** from the second, **3** from the third, up to **12** from the twelfth sack. Therefore, you have $1 + 2 + 3 + 4 + 5 + 6 + 7 + 8 + 9 + 10 + 11 + 12$, or **78** coins. Seventy-eight coins, if they are not counterfeit, *should* weigh 78 pounds. So, weigh your coins, and see *how many ounces you are over* that 78 pounds. The number of ounces you are over corresponds to the number of extra coins you have which weigh that extra ounce. For example, if it weighs 11 extra ounces, you must have 11 coins which weigh 1 pound 1 ounce. They came from sack 11, so that must be the bag containing the counterfeit coins. If it had weighed 7 extra ounces, that would have meant 7 coins at "1 ounce over," so that would have meant sack 7 contained the counterfeits, and so forth.

Chapter 10

Problems Involving Problem Solving Skills Other Than Arithmetic

There are many "number problems" that can be solved very easily *without* using any (or at least not very much) arithmetic. In these problems, we rely more on logic and thinking. Even when or if arithmetic is still involved, it is generally much *simpler* arithmetic.

We talked earlier about using patterns, working backward, **(educated)** guess and check, and known arithmetic shortcuts involving things like divisibility. Let's go into these strategies in greater depth.

There is a story told about Carl Friedrich Gauss, a famous 18th century German mathematician, that when he was in kindergarten, his teacher had to leave the room for a few moments, and wanted to keep the class **very** busy. So, she asked the children to add all the numbers from 1 to 100. The children started working, and she thought she had found a successful exercise. Much to her surprise, however, little Carl had his hand in the air even **before** she could leave the room! Not only had he completed the assignment, but he did, in fact, have the correct answer!

His "secret" was not a calculator, nor was it an ability to truly add a column of 100 numbers so quickly. Rather, he found a "pattern." Knowing that both the associative and commutative properties hold for addition (though I seriously doubt he knew the names, or even that they were properties), he realized that he could add the numbers in any order. Looking at the column will make it easier to understand what he did.

1
2
3
4
5
.

.

.

47
48
49
50
51
52
53
.

.

.

95
96
97
98
99
100

He saw that if he added the first and next to last numbers, he got a sum of 100. If he added the second and next to last numbers, he again got a sum of 100. In fact, every number on his list could be paired with another to give a sum of 100. How many of these pairs did he have?

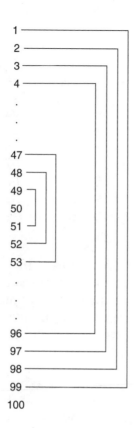

There were 49 of these pairs. (Each number from 1 through 49 could be paired with another to give a sum of 100.) That comes to 4900. But there was still a 100 unaccounted for (taking the sum to 5000), as well as a 50. This brought the final sum to **5050.**

There have been variations on this idea over the years. For example, you can add the first and last, getting a sum of 101, and the second and next to last, getting a sum of 101, and so forth all the way through to adding 50 and 51. The final result, therefore, must be 101×50 or 5050. This method **is** faster than the first, but remember that little Carl was only in kindergarten and might not have been able to multiply!

Another variation involves less strategy and more *observation*. If you start adding but break up the column so that it is a little easier to work with, you would get something like this:

1	11	21	31	41	51	61	71	81	91
2	12	22	32	42	52	62	72	82	92
3	13	23	33	43	53	63	73	83	93
4	14	24	34	44	54	64	74	84	94
5	15	25	35	45	55	65	75	85	95
6	16	26	36	46	56	66	76	86	96
7	17	27	37	47	57	67	77	87	97
8	18	28	38	48	58	68	78	88	98
9	19	29	39	49	59	69	79	89	99
10	20	30	40	50	60	70	80	90	100
55	**155**	**255**	**355**						

What do you think the remaining sums would be? AND, if we
add those **sums,** we get (of course) **5050.**

We could even use the pattern approach on problems involving exponents, on problems involving geometry, on problems involving fractions, and many other topics as well.

MODEL PROBLEM I

Find the units digit if 2 is raised to the 74th power. Start slowly! Don't expect to be able to get the final answer immediately! Find the powers one by one... and then see if you can find a pattern.

$$2^1 = 2$$
$$2^2 = 4$$
$$2^3 = 8$$
$$2^4 = 16$$
$$2^5 = 32$$
$$2^6 = 64$$
$$2^7 = 128$$

Do you see the pattern? The units' digits of 2, 4, 8, and 6 have begun to repeat. The cycle is 4. Now we just have to see where 74 would fall into the cycle. The greatest multiple of 4 less than 74 is 72. We then have to move 2 more places to find where the 74th term would be. **The units' digit,** therefore, **is 4.**

MODEL PROBLEM 2

How many diagonals does a nine-sided polygon have? You could try to draw the polygon and count, BUT there aren't too many people who can draw a nine-sided polygon (a nonagon). So, let's start slowly. Start with a triangle, draw the polygon, and determine the number of diagonals. Keep increasing the number of sides, making sure to draw the figure, and keep a table of the number of sides and the number of diagonals.

[Remember that a diagonal is any line segment drawn from any vertex of a polygon to another NONADJACENT vertex of the polygon. Also remember that when you are counting diagonals, do not "recount" a diagonal going in the "opposite direction" ($\overline{AC}$ is the same diagonal as $\overline{CA}$).]

	NUMBER OF SIDES	NUMBER OF DIAGONALS
	3	0
	4	2
	5	5
	6	9

By this time, hopefully, you will have found a pattern, because in all likelihood you will not be able to draw a 7-sided polygon easily.

Look at the table. Clearly the number of diagonals is not a multiple of the number of sides. But since we have increased the number of sides one by one, it is easy to examine what has happened to the number of diagonals with each such increase. Although there has not been a *constant* increase, there has been a **steady** increase. The first time, the number of diagonals increased by 2 (from 0 to 2), then by 3 (from 2 to 5), and then by 4 (from 5 to 9). If this *pattern* is to continue, the next time the number of diagonals must increase *by 5, from 9 to 14.* Therefore, a 7-sided polygon must have 14 diagonals. Following this approach, an 8-sided polygon must have $14 + 6$, or *20* diagonals, and a **9-sided polygon** must have $20 + 7$ or **27 diagonals.**

Now a pattern approach works very nicely with a problem like this IF you're not being asked to find the number of diagonals of a polygon of *100* sides. You COULD do the problem following what we did for this problem, BUT it would become a long and tedious process. You COULD look at the multiples (there is a pattern to that), but in this case it would be even easier to use some LOGIC and try to figure out a formula. You think about it, and I'll show it to you in the SOLUTIONS section of the chapter.

MODEL PROBLEM 3

What is the sum of the fractions indicated below?

$$\frac{1}{1 \times 2} + \frac{1}{2 \times 3} + \frac{1}{3 \times 4} + \frac{1}{4 \times 5} + \frac{1}{5 \times 6} + \cdots + \frac{1}{9 \times 10}$$

If you are anything like my students, you very well may have what I always term an "allergy" to fractions. Somehow, just seeing a fraction makes so many students feel that they will not be able to do the problem. Unfortunately, *if you do a good enough job telling yourself you cannot do something, you won't be able to.* First of all, fractions are a lot easier than you think, and secondly, in this problem, let's see what we have BEFORE we start worrying.

$$\frac{1}{2} + \frac{1}{6} + \frac{1}{12} + \frac{1}{20} + \frac{1}{30} + \cdots + \frac{1}{90}$$

Considering the fact that most students don't even like finding the common denominator for two fractions, finding one for 9 fractions would not be a particularly "fun" way to go. So, let's try to look for a pattern. Maybe we won't have to actually add ALL the fractions.

If we consider just the first fraction, we, of course, get **1/2.**

If we add the first two fractions, we have $1/2 + 1/6$, which equals $3/6 + 1/6$ or $4/6$ or **2/3.**

If we add the third fraction, we have $2/3 + 1/12$, which equals $8/12 + 1/12$ or $9/12$ or **3/4.**

If we add the fourth fraction, we have $3/4 + 1/20$, which equals $15/20 + 1/20$ or $16/20$ or **4/5.**

Look at the sums we have so far, and see if the numbers involved look at all familiar. Have we seen these numbers anywhere in the problem? I'll give you a hint. Look at the denominators in the original problem. Each was a product. The *factors* in the first denominator were *1* and *2*. Notice that the "sum" of the first fraction is 1/2. The *factors* in the second denominator were *2* and *3*. The sum of the first two fractions was *2/3*. The *factors* in the third denominator were *3* and *4*. The sum through the third fraction was *3/4*. The sum through the fourth fraction was *4/5,* and the *factors* in the denominator were *4* and *5*. From this pattern, we can tell that the **sum of the 9 fractions given to us will be** a fraction whose numerator and denominator are the two factors, or **9/10.**

In Chapter 1, we discussed, briefly, the "shortcut" involving the rule of divisibility by 3 and 9. Do you remember it? While you are trying to think of it, I'll just quickly mention some others. We know, of course, that all even numbers are divisible by 2. And these even numbers are, of course, easily identifiable by the fact that the units' digit is even, or as some people like to think of it, they "end" with an even number. But can we just as easily determine divisibility by 4? All you have to do is to examine the ten's AND units' digits (the "last two" digits). As long as that 2-digit number is divisible by 4, the original number must be divisible by 4.

For example, just looking at that one kind of backward, 24 is a number divisible by 4. If 24 represented the "last two" digits of ANY number, that number would be divisible by 4. So that 124 1,224 12,324 123,424 1,234,524 12,345,624 are all examples of numbers divisible by 4. Do you have any ideas about how you might test for divisibility by 8? (Answer in the answer section.)

Have you remembered the test for divisibility for 3 and the one for 9 yet? Very simply, if the sum of the digits of the number is divisible by 3, the number is divisible by 3, and if the sum of the digits of the number is divisible by 9, the number is divisible by 9.

Now, how can we use this knowledge to help us solve problems?

MODEL PROBLEM 4

A three-digit numeral in base 10 represents a multiple of 9. After crossing out one of the digits, 4 and 8 are left. What digit was crossed out? You *could* do this problem with arithmetic and trial and error (and hope that you don't get careless with your division), or you could remember the test for divisibility by 9, and add the two digits you have. When you add 4 and 8, you get 12. The next multiple of 9 is 18. What do you have to add to 12 to get 18? **6** is the **digit that was crossed out.**

What else do we know about numbers divisible by a specific number? Let's make a list of those divisible by 2, by 3, by 4, and by 5. (Look for a pattern as we go.)

NUMBERS DIVISIBLE BY 2: 2, 4, 6, 8, 10, 12, 14, 16, 18, 20, 22, 24, ...

NUMBERS DIVISIBLE BY 3: 3, 6, 9, 12, 15, 18, 21, 24, 27, 30, 33, 36, ...

NUMBERS DIVISIBLE BY 4: 4, 8, 12, 16, 20, 24, 28, 32, 36, 40, 44, 48, ...

NUMBERS DIVISIBLE BY 5: 5, 10, 15, 20, 25, 30, 35, 40, 45, 50, 55, 60, ...

Even without knowing a "times table," you should be able to figure out the next number in each sequence. In addition, look at *how many* numbers there are in each group.

- In the first group, where we counted to 24, how many numbers were there divisible by 2? *There were 12.*
- In the second group, where we counted to 36, how many numbers were there divisible by 3? *There were 12.*
- In the third group, where we counted to 48, how many numbers were there divisible by 4? *There were 12.*
- In the fourth group, where we counted to 60, how many numbers were there divisible by 5? *There were 12.*

Is there a relationship between *how many* numbers satisfied the conditions we established, and the conditions? Notice that, in each case, if we divide the number of numbers considered by the number by which they are to be divisible, we will find exactly how many there should be. Simply put, that is because every other number is divisible by 2; every third number is divisible by 3; every fourth number is divisible by 4, and so forth.

Can we use this knowledge to help us with "number" problems? Of course.

MODEL PROBLEM 5

How many numbers are there between 250 and 550 which are divisible by 3?

You could start listing the numbers (and you don't even have to do any division; remember that the sum of the digits must be divisible by 3, so *252* would be the first in this set to satisfy the condition), or you could simply use some logic. There are 300 numbers between 250 and 550. Every third number is divisible by 3, so divide 300 by 3, and we find that **there are 100 numbers divisible by 3 between 250 and 550.**

Another type of problem involves reasoning far more than arithmetic. While it is true that you need arithmetic to ultimately do the problem, all the arithmetic and mathematical skills in the world *won't help if you don't think*!

MODEL PROBLEM 6

It costs 96 cents to cut a piece of pipe into 3 pieces. How much should it cost to cut it into 6 pieces?

Don't jump to the obvious conclusion! The answer is NOT $1.92. Before we discuss the solution, "draw" the piece of pipe on your paper. Now draw in the "cuts." HOW MANY CUTS DID YOU HAVE TO MAKE TO FORM THE **3** PIECES? Your diagram should look like this:

YOU ONLY HAD TO MAKE **2** CUTS! The second cut formed the second and third pieces. That means that the 96 cents was the cost for **2 cuts**. EACH CUT COST 48 CENTS. Now, HOW MANY CUTS WOULD YOU HAVE TO MAKE IF YOU WANTED **6** PIECES? Remember that one cut forms the last two pieces! YOU WOULD NEED **5** CUTS.

Now, just multiply 48 × 5, and we find that it would cost **$2.40.**

Practice Problems

1. Find the units' digit if 2,345,432 is raised to the 154th power.
2. Find the units' digit of 3 to the 20th power.
3. Find the remainder if 3 to the 20th power is divided by 7.
4. Find the digit in the 57th place if 4/7 is expressed in decimal form.
5. How many diagonals are there in a dodecagon? (If you're not sure how many sides there are in a dodecagon, look it up!)
6. If 123,456,7A8 is to be divisible by 9, find the value of A.
7. How many multiples of 5 are there between 103 and 203?
8. How many integers are there between 199 and 499 which, when divided by 3, leave a remainder of 2?
9. How many integers between 100 and 1000 are divisible by 7 or by 11?
10. The cost of cutting a pipe into 4 pieces is $4.20. What should be the cost of cutting it into 8 pieces?

11. (This is a problem from the NYC Math Team Competitions that has given students trouble for years. BE CAREFUL!) A man has 3 loaves of bread, and another has 5 loaves of bread. They agree to share their bread with a third man, who had none, so that each will have an equal share. For this, the third man pays 80 cents. How much of the 80 cents should the man who had the 5 loaves receive?

12. (Another Math Team Competition problem, given in 1976, the year we celebrated the Bicentennial) The number represented by 1**aa**6 is divisible by 37 and by 3. Find the value of **a**.

Solutions to the Practice Problems

1. We are asked to find the units' digit of $(2,345,432)^{154}$. First of all, we have to remember that the only thing that will affect the units' digit is the units' digit of the original number. The problem now becomes one of finding the units' digit of 2^{154}. Let's look for that pattern. (Multiplication is no big deal, so the only reason a question like this would be asked is to see if you know WHAT to do.)

$$2^1 = \mathbf{2}$$
$$2^2 = \mathbf{4}$$
$$2^3 = \mathbf{8}$$
$$2^4 = \mathbf{16}$$
$$2^5 = \mathbf{32}$$

The units' digit of 2 has reappeared, and that means that the cycle has started once again. How many different units' digits are there in the cycle? There are **4**. Therefore, we have to divide 154 by 4, and find where the 154th term would appear in the cycle. If we divide 154 by 4, we get 38 **with a remainder of 2**. Therefore, find the second term in the sequence, which means that the units' digit is the one in the "second position," or **4**.

2. To find the units' digit of 3^{20}, we will find the various powers of 3.

$$3^1 = \mathbf{3}$$
$$3^2 = \mathbf{9}$$
$$3^3 = \mathbf{27}$$
$$3^4 = \mathbf{81}$$
$$3^5 = \mathbf{243}$$

Notice that the units' digit has started to recur. Again we have a cycle of 4. (This is a *coincidence*, NOT A PATTERN!) Where would the 20th term appear? Since 20 is divisible by 4, 3^{20} would be in the "fourth position." Therefore, the **units' digit of 3^{20} is 1.**

3. We might as well use the list of powers of 3 that we developed for the last problem. (We certainly do not want to have to first find the value of 3^{20} and then do our division.) For this problem, however, we have to add a column where we can list the *remainder* when that power is divided by 7.

    $3^1 = 3$. If we divide 3 by 7, we get a quotient of 0, *and a remainder of* <u>3.</u>

    $3^2 = 9$. If we divide 9 by 7, we get a quotient of 1, *and a remainder of* <u>2.</u>

    $3^3 = 27$. If we divide 27 by 7, we get a quotient of 3, *and a remainder of* <u>6.</u>

    $3^4 = 81$. If we divide 81 by 7, we get a quotient of 11, *and a remainder of* <u>4.</u>

    $3^5 = 243$. If we divide 243 by 7, we get a quotient of 34, *and a remainder of* <u>5.</u>

    (No pattern yet, so we have to keep going.)

    $3^6 = 729$. If we divide 729 by 7, we get a quotient of 104, *and a remainder of* <u>1.</u>

    $3^7 = 2187$. If we divide 2187 by 7, we get a quotient of 312, *and a remainder of* <u>**3.**</u>

    Now we have our pattern. This cycle is one of 6 terms. The 20th power of 3 would be the 20th term, which puts it in the *second position* (18 is our multiple of 6, and we need 2 more for 20). Therefore, **the remainder is 2.**

4. We can express any fraction as a decimal by dividing the numerator by the denominator. If we divide 4 by 7, we get $.\overline{571428}$. If we want to know what digit will be in the 57th place, just figure out where the 57th term would be in the cycle. We see that we have a cycle of 6. The greatest multiple of 6 which is less than 57 is 54. There is a remainder of 3. Therefore, the **57th digit** will be the same as the third one in the cycle, or **1.** (Incidentally, "FYI,"

$$1/7 = .\overline{142857}$$
$$2/7 = .\overline{285714}$$
$$3/7 = .\overline{428571}$$

$$4/7 = .\overline{571428}$$
$$5/7 = .\overline{714285}$$
$$6/7 = .\overline{857142})$$

Do you notice anything unusual about these decimals? Notice that not only is each one made up of the same 6 digits in the repetend, but those digits are in the same order; there is simply a different "starting point." Can you find any other fractions for which this is true?

5. First of all, a dodecagon has 12 sides. Now, having said that, we have two choices. We can continue with the pattern approach we considered earlier, in the model problem, or we can try to find a formula. Let's start with the pattern. In the earlier problem, we had already found this pattern:

NUMBER OF SIDES		NUMBER OF DIAGONALS
3		0
4		2
5		5
6		9
7		14
8		20
9		27

Continuing,

10	(we must now add 8)	35
11	(we must now add 9)	44
12	(we must now add 10)	**54**

As we can see, although it is now difficult, it can become rather tedious to work out this problem as the number of sides increases, if we have to rely on the pattern. I asked you earlier to see if you could come up with a formula. Were you able to? Think of how many diagonals you can draw from ANY vertex of any polygon. Since a diagonal is a line segment drawn from any vertex to any nonadjacent vertex, any given vertex can be joined to all but three of the vertices: itself, and the two on either side of it. Therefore, if the polygon has n sides (and, therefore, n vertices) there will be

$n - 3$ diagonals that can be drawn from that vertex. Since there are n vertices, there will be $n(n - 3)$ diagonals. However, since each diagonal can be "drawn twice" (in either direction), we have to divide by 2. The formula, then, is $\frac{n(n-3)}{2}$. Just to prove that we are correct, let's test the formula on the problem we just did. If there are 12 sides, $n = 12$ and $n - 3 = 9$. $(12)(9) = 108$. $108/2 = 54$, just as we found with our pattern.

6. We could use trial and error, substituting 0 through 9 for A, to see which one is correct, but it will be much easier if we remember that the sum of the digits must be divisible by 9. Adding what we have, $1 + 2 + 3 + 4 + 5 + 6 + 7 + 8 = 36$. Since 36 IS a multiple of 9, A can equal **0**. OR, since 45 is a multiple of 9, A could equal **9** also.

7. Again, you can do this problem "the long way," and list all the numbers satisfying the required condition, OR, we can simply figure that there are 100 numbers between 103 and 203. Every fifth one is a multiple of 5. Therefore, **there are 20 numbers between 103 and 203, which are multiples of 5.**

8. This one may look more complicated, but in reality it is the same as the others we have done. Even though we are looking for a remainder this time, it is STILL every third number that will satisfy the condition. (If you look at a list of numbers, you will see that you simply have to count 2 past each of the multiples of 3, but that still means every third one!) So, $499 - 199 = 300$. $300/3 = 100$. **There are 100 numbers satisfying this condition.**

9. There are 900 numbers between 100 and 1000. Every seventh one is divisible by 7. **$900/7 = 128$.** Every eleventh one is divisible by 11. **$900/11 = 81$.** Since either condition is acceptable, $128 + 81 = $**209.** **However, this is not the answer.** The first group accounts for *all the numbers divisible by 7,* and the second group accounts for *all the numbers divisible by 11.* We certainly have accounted for *all of* them, BUT HAVE WE ACCOUNTED FOR ANY OF THEM MORE THAN ONCE? *We have, if there are numbers that will are divisible by 7* **and** *by 11.* There are such numbers. They are *divisible by 77.* How many of these numbers are there? Just divide 900 by 77. $900/77 = 11$. We therefore have to subtract 11 from the 209 we got earlier, and $209 - 11 = 198$. **There are 198 integers between 100 and 1000, which are divisible by 7 or by 11.**

10. *Four pieces* would be formed by making *3 cuts.* Therefore, $4.20 for 3 cuts means that each cut costs $1.40. To cut the pipe into 8 pieces, we need 7 cuts. The cost of the 7 cuts would be **$7 \times $1.40,** or **$9.80.**

122

11. First, let me say, THE ANSWER IS *NOT* 50 CENTS. But, if that is the answer you got, you are not alone! In fact, that is the answer that most people get. The "catch" is that the third man is NOT buying all the bread. He is <u>sharing</u> it with the other two men. To make the explanation a little simpler, let's call the two men with the bread A and B, and the man with the money C. If A has 3 loaves, and B has 5 loaves, there will be a total of 8 loaves of bread. But you probably figured that when you arrived at the *50 cents*. But, as stated earlier, C is NOT buying the 8 loaves. C is *sharing* the 8 loaves *with* the other two men. That means that each of the men will ultimately have 8/3 or $2\frac{2}{3}$ loaves. If A has 3 loaves now, he must GIVE UP 1/3 of a loaf. If B has 5 loaves now, he must give up $2\frac{1}{3}$, or 7/3, loaves. Therefore, although the bread they HAVE is in the ratio of 3 to 5, the BREAD THEY ARE SELLING TO THE THIRD MAN is in the ratio of **1 to 7.** Therefore, they must be **paid** in the ratio of **1 to 7,** and the man with the 5 loaves originally will get **70 cents.**

12. Again, you can try trial and error, guess and check, or whatever you want to call it. But that can be time-consuming. Think, instead, of what divisibility by two numbers would have to mean. For a number to be divisible by both 37 and 3, it must be divisible by 37 × 3. 37 × 3 = 111. Therefore, our number is divisible by 111. Since we know that 10 × 111 = 1110 and 20 × 111 = 2220, we know that the multiple must be between 10 and 20. In addition, if we need a units' digit of 6, we have to multiply by a factor with a units' digit of 6. The only number between 10 and 20 with a units' digit of 6 is 16. Therefore, the only *possibility* (and we would still have to check to make sure that the product was of the correct form) is 111 × 16. If we multiply 111 by 16, we get 1**776.** Notice, as required, the second and third digits are the same. (And, perhaps, the **1776** was in honor of the Bicentennial.) By the way, even if you did not see this shortcut, you should not have had to try EVERY value of **a** from 0 through 9. Since the number must be divisible by 3, the sum of the digits must be a multiple of 3.

If **a** = **0**, 1 + 0 + 0 + 6 = **7**, which is not a multiple of 3.
If **a** = **1**, 1 + 1 + 1 + 6 = **9**, which IS a multiple of 3.
If **a** = **2**, 1 + 2 + 2 + 6 = **11**, which is not a multiple of 3.
If **a** = **3**, 1 + 3 + 3 + 6 = **13**, which is not a multiple of 3.
If **a** = **4**, 1 + 4 + 4 + 6 = **15**, which IS a multiple of 3.
If **a** = **5**, 1 + 5 + 5 + 6 = **17**, which is not a multiple of 3.
If **a** = **6**, 1 + 6 + 6 + 6 = **19**, which is not a multiple of 3.

If **a** = **7**, $1 + 7 + 7 + 6 = 21$, which IS a multiple of 3.
If **a** = **8**, $1 + 8 + 8 + 6 = 23$, which is not a multiple of 3.
If **a** = **9**, $1 + 9 + 9 + 6 = 25$, which is not a multiple of 3.

Therefore, the only value of **a** you would have to check would be 1 or 4 or 7. *Guess and check* is a wonderful method, but always try to make an **educated** *guess*! AND, of course, <u>make sure you answer the question</u>! The problem asked you to "find the value of **a**." So, **a** = **7**.

Chapter 11

Some Mathematical Curiosities and Other Fun Stuff

As the title suggests, this chapter will just be devoted to some "unusual" facts that might help you amaze your friends and classmates.

For example, take the number 12,345,679. (Be very careful when you write it or enter it into the calculator; *there is no 8*.) Now, multiply it by 9. Notice anything unusual? Now try multiplying by 18. Then multiply it by 27. What do you notice about the products?

If you did the multiplication correctly, the products you should have gotten were:

$$12,345,679 \times 9 = 111,111,111$$
$$12,345,679 \times 18 = 222,222,222$$
$$12,345,679 \times 27 = 333,333,333$$

Quite obviously, these are very special products. But now the question is WHY? And, CAN WE PREDICT OTHER PRODUCTS? DO WE HAVE TO MULTIPLY BY "SPECIAL NUMBERS"?

Look carefully at the three problems we have done. In each, of course, we have a factor of 12,345,679, but *let's examine the other factors*. Is there anything that 9, 18, and 27 have in common?

I hope you saw that all are multiples of 9 (or, in other words, divisible by 9). If we divided 9 by 9, we get 1. Notice that when we multiplied our "magic number" by 9, we got all 1's. Dividing 18 by 9, we get 2. When we multiplied our "magic number" by 18, we got all 2's. Does the pattern continue with 27?

What product would we get if we multiply 12,345,679 by 54?

By what number do you think we would have to multiply if we wanted a product consisting of all 5's?

Answers to these questions later.

This "trick" even works in other bases. If you use *all the permissible digits in the base EXCEPT FOR THE NEXT TO LAST ONE* and then *multiply by MULTIPLES OF ONE LESS THAN THE BASE*, the same thing will happen!

Here's another "curiosity." Find each of the following products:

12	23	35	46	54	68	71	87	99
×18	×27	×35	×44	×56	×62	×79	×83	×91

Hopefully, you got:

$$12 \times 18 = 216$$
$$23 \times 27 = 621$$
$$35 \times 35 = 1225$$
$$46 \times 44 = 2024$$
$$54 \times 56 = 3024$$
$$68 \times 62 = 4216$$
$$71 \times 79 = 5609$$
$$87 \times 83 = 7221$$
$$99 \times 91 = 9009$$

Now examine each of those products carefully. Notice anything unusual?

$$1\underline{2} \times 1\underline{8} = 2\underline{16}$$
$$2\underline{3} \times 2\underline{7} = 6\underline{21}$$
$$3\underline{5} \times 3\underline{5} = 12\underline{25}$$
$$4\underline{6} \times 4\underline{4} = 20\underline{24}$$

$$54 \times 56 = 30\underline{24}$$
$$68 \times 62 = 42\underline{16}$$
$$71 \times 79 = 56\underline{09}$$
$$87 \times 83 = 72\underline{21}$$
$$99 \times 91 = 90\underline{09}$$

How often have you been able to multiply two 2-digit numbers and just write down the product of the units' digits as the "last two" digits of the product?

What about the rest of the product? Is there anything "magical" about that? Look carefully at the problems and at the products. Notice that in each of the multiplication problems the factors have the same tens' digits. If we then "make believe" that one of them is one more than it is, and we then multiply it by the other, we get the "first two" digits of the product. That is,

$$\text{"2"} \times 1 = 2$$
$$\text{"3"} \times 2 = 6$$
$$\text{"4"} \times 3 = 12$$
$$\text{"5"} \times 4 = 20$$
$$\text{"6"} \times 5 = 30$$
$$\text{"7"} \times 6 = 42$$
$$\text{"8"} \times 7 = 56$$
$$\text{"9"} \times 8 = 72$$
$$\text{"10"} \times 9 = 90$$

Obviously, though, this "shortcut" cannot work with all products of all 2-digit numbers. We have already seen that the *tens' digits are the same*. What about the *units' digits*? Do you see anything special about:

2 & 8

3 & 7

5 & 5

6 & 4

4 & 6

8 & 2

1 & 9

7 & 3

9 & 1

In each case, the *sum of the numbers is 10.*

So, summing up, if you are multiplying two 2-digit numbers where the tens' digits are the same, and the sum of the units' digits is 10, just write down the product of the units' digits as the "last two" digits of the product (if the product is 9, you will have to write 09), and then increase one of the tens' digits by 1 and multiply it by the other. That product becomes the "first two" digits of the product.

Another "shortcut" involves multiplication by 11.

Find these products:

24	231	3421	43,212	321,423
×11	×11	×11	×11	×11

Did you get these products?:

$$24 \times 11 = 264$$
$$231 \times 11 = 2541$$
$$3421 \times 11 = 37,631$$
$$43,212 \times 11 = 475,332$$
$$321,423 \times 11 = 3,535,653$$

Notice the first and last digits of the first factor and of the product in each case:

$$\mathbf{24} \times 11 = \mathbf{264}$$
$$\mathbf{231} \times 11 = \mathbf{2541}$$
$$\mathbf{3421} \times 11 = \mathbf{37},631$$
$$\mathbf{43},212 \times 11 = \mathbf{475},332$$
$$\mathbf{321},423 \times 11 = \mathbf{3},535,653$$

In the first product, you might also notice that $2 + 4 = $ **6** (the middle digit of the product). Look a little more closely, and you will notice in the second problem (working from the right as we would when we multiply) that $1 + 3 = $ **4,** and that $3 + 2 = $ **5.**

In the third problem:

$$1 + 2 = \textbf{3,} \qquad 2 + 4 = \textbf{6,} \qquad 4 + 3 = \textbf{7.}$$

Check this pattern in each of the other problems.

What do you think happens if the sum is such that you have to "carry"? Do exactly what you would do under normal circumstances. Put down the units' digit of that sum and "carry," making sure to include that in your next sum. If you have to "carry" after your last sum, just increase that digit by 1 before you bring it down. Let's "sum up" with a problem:

$$\begin{array}{r} 263{,}547{,}485 \\ \times\,11 \\ \hline \end{array}$$

Starting from the right, bring down the 5:	5
Next add $5 + 8$. That's 13. Write down the 3 and carry 1:	35
Next add $8 + 4$ and carry 1. That's 13. Write 3, carry 1:	335
Next add $4 + 7$ and carry 1. That's 12. Write 2, carry 1:	2335
Next add $7 + 4$ and carry 1. That's 12. Write 2, carry 1:	22,335
Next add $4 + 5$ and carry 1. That's 10. Write 0, carry 1:	022,335
Next add $5 + 3$ and carry 1. That's 9.	9,022,335
Next add $3 + 6$. That's 9.	99,022,335
Next add $6 + 2$. That's 8.	899,022,335
Finally, bring down the 2.	2,899,022,335

See if you can figure out WHY this works.

Have you ever heard or sung the song "The Twelve Days of Christmas"? Did you ever stop to think of just HOW MANY gifts my "true love gave to me"? If you don't remember all the lyrics, don't worry. It really doesn't matter whether you remember how many "lords were leaping" or how many "ladies were dancing," as long as you remember the basic pattern described in the song:

(By the way, the "partridge in a pear tree" counts as ONE gift.)

I'll get you started. Then it will be your job to find out HOW MANY GIFTS WERE GIVEN over the course of the twelve days.

On the first day of Christmas
  my true love gave to me:

a partridge in a pear tree
<hr>
1 gift

On the second day of Christmas
  my true love gave to me:

2 turtle doves   and
a partridge in a pear tree
<hr>
3 gifts

On the third day of Christmas
  my true love gave to me:

3 French hens
2 turtle doves   and
a partridge in a pear tree
<hr>
6 gifts

On the fourth day of Christmas
  my true love gave to me:

4 calling birds
3 French hens
2 turtle doves   and
a partridge in a pear tree
<hr>
10 gifts

Now see if you can find the total number of gifts. Care to make a guess first? (I'll let you know whether or not you were correct later.)

There's also a great card trick with which you can amaze your friends. And there really is no magic to it at all. It's all mathematical!

Take a deck of cards, and put down (face up) 3 cards next to one another. (They will be the tops of three columns.) Next put down 3 more, touching but not covering the first three. Continue this until you have 7 rows in your 3 columns.

Card 1	Card 2	Card 3
Card 4	Card 5	Card 6
Card 7	Card 8	Card 9
Card 10	Card 11	Card 12
Card 13	Card 14	Card 15
Card 16	Card 17	Card 18
Card 19	Card 20	Card 21

Ask your friends to "pick a card" but NOT tell you what it is. Have them, however, tell you WHICH COLUMN it is in. Pick up one of the other columns, then the "selected column," and then the other column. Put the cards down again in the same fashion as you did originally. Have your friends look for the card they originally chose. Again, have them tell you WHICH COLUMN it is in. Pick up the columns using the method described earlier. Then put the cards down (in the 3 columns, working ACROSS) a third and last time. Again, have your friends tell you the column in which you would find the card. And, one last time, pick up the cards, column by column, with the selected column in the middle.

What your friends do not know, and there is no reason to tell them, is that by making sure the column containing the card is in the middle each time, and then putting the cards out going across, the card they have selected is now the MIDDLE CARD in the set of 21 cards. In other words, it is the 11th card. You could in fact give it to them right now, but here's how to make it look even more amazing. Place 4 cards down (FACE DOWN) one at a time, in a pinwheel fashion, as shown below. Keep track, in your mind, where you place the 11th card. Remember, that's the one your friends have chosen. The cards should now look like this:

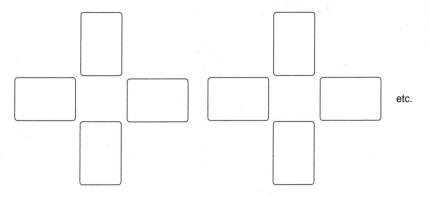

etc.

(You'll have one left over, but just put it to the side.) Now, the object is to, "miraculously," have your friends' card appear at the end of the trick. Tell your friends to pick 2 packs (2 pinwheels). *If they pick the pack that contains the 11th card*, <u>get rid of the packs they have NOT selected</u>. *If they do NOT pick the pack that contains the 11th card*, <u>get rid of the packs they HAVE selected</u>. If you have 3 packs left, again have them select two, and follow the same procedure as above. If you have 2 packs left, ask them to just pick one. Either way, just remember that you have to KEEP THE PACK that CONTAINS THE 11TH CARD. Now have them pick 2 cards, again keeping or getting rid of the cards based on KEEPING THE 11TH CARD. Ultimately, you will be down to one card. Have your friends turn it over. WATCH THEIR SURPRISE WHEN THE CARD THEY TURN OVER IS THE ONE THEY CHOSE. They think it's **magic;** you know it's **mathematics.**

Chapter 12

Miscellaneous Problem Drill

1. Julian wishes to put a fence around his backyard. The property to be enclosed is 60 feet by 30 feet. He is going to use boards which will be placed vertically. Each board is 6 inches wide. How many of these boards will he need?

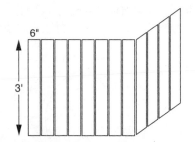

2. If the fence is to be 3 feet high, and 1 gallon of paint will cover 24 square feet, how many gallons of paint will Julian need to cover the fence?

3. How many square feet of carpet will Ms. Fletcher need to carpet her living room, dining room, and hallway if the apartment layout looks like the diagram below?

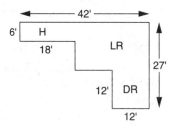

4. If the carpeting Ms. Fletcher has chosen costs $22.50 a square **yard,** how much will it cost her to carpet the area?

5. Columbo's Pizza advertises a 14-inch pizza for $11 and an "individual" 7-inch pizza for $5.50. Which would be the better buy: one 14-inch pizza or two 7-inch pizzas?

6. Mr. Dowling decided he wanted a circular garden in his back yard. The diameter of the garden would be 20 feet. First, he will encircle the garden with a wire mesh fence. How many feet of fencing would he have to buy? If the fencing costs $2.50 a foot, how much will it cost him for his fence? After he encloses the garden, he will be planting. If a bag of seed will cover 150 square feet, how many bags of seed will Mr. Dowling need? If the seed costs $11 per bag, how much will it cost him?

7. Ms. King wants to have a new finish put on the silver frame in which she has her wedding picture. The frame is circular and is 3 inches wide. The frame looks like the diagram below. She has been told that the new finish will cost her $0.75 per square inch. What would be the cost?

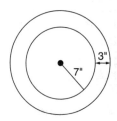

8. What is the measure of the vertex angle of an isosceles triangle with a base angle whose measure is 57°?

9. Find the measure of a base angle of an isosceles triangle if the measure of the vertex angle is 74°.

10. If the measures of the angles of a triangle are in the ratio of 1 : 5 : 6, find the measure of each angle. What KIND of triangle is it?

11. Michael's kite is caught in a tree. Attached to the kite was a 100-foot string. With the string stretched out straight, Michael can stand 28 feet from the foot of the tree. What is the height of the tree?

12. If a triangle has sides of 8 inches, 33 inches, and 34 inches, will it be a right triangle?

13. In the Central Valley MS there is a rule that no activity can be held unless 90% of the eligible students choose to participate. If there are 250 students in the eighth grade, how many would have to agree to go on the senior trip in order for it to be held? So far, 191 students have signed up. Is that enough? If not, how many more students have to agree to participate?

14. Ms. Jackson just ordered a car where she is required to give the dealer a deposit of 15%. If she gave the dealer $5250, what is the total price of the car? How much will she have to give the dealer when she is ready to pick up the car?

15. Stephanie was paid a commission of $4000 last week. She put $2500 of that in the bank. What percent of her commission did she **keep?**

16. Marsha has a lot of great recipes, but unfortunately they all serve 4 people and she has to serve 6. If her recipe for a special pudding calls for 2/3 cup of milk, how many cups of milk will she actually need?

17. Susan decided to check her map before leaving on a long trip, and found that, on the map, the distance between home and her destination was $4\frac{1}{2}$ inches. If the scale on the map is 2 inches represents 55 miles, how far does she actually have to travel?

18. Ed found that he could buy a $4\frac{1}{2}$-gallon can of paint for $23.50. The store also had a 10-gallon can of the same paint for $52.50. Which is the better buy?

19. If Ed needs 45 gallons of the paint mentioned in Problem 18, what would it cost him IF HE BUYS THE SIZE, WHICH IS THE BETTER BUY?

20. Ken had test scores of 93, 92, 94, 98, and 88. What would he need on his next test in order to have an average of 94? Would Ken be able to achieve an average of 95 with his next test? Why or why not?

21. Given the following sets of averages and community service hours for the last three years, determine who would be eligible for the Honor Society if the requirements are an

average of 90 and an **average of 30 hours of community service:**

NAME	AVERAGE EACH YEAR	NUMBER OF HOURS OF COMMUNITY SERVICE
Amanda	93, 88, 97, 80	29, 35, 34, 36
Alice	93, 99, 99, 68	34, 34, 34, 19
Annie	87, 87, 87, 100	31, 31, 31, 26

22. Find the mean, median, mode, and range for the following set of scores:

97	93	92	88	97
97	87	99	92	86
93	91	97	95	89

23. I am thinking of two numbers. Their sum is 83, and their difference is 15. Find the numbers.

24. Find the smallest number such that when it is divided by 2, there is a remainder of 1; when it is divided by 3, there is a remainder of 2; when it is divided by 4, there is a remainder of 3; and when it is divided by 5, there is a remainder of 4.

25. Two numbers are in the ratio of 7 : 5. Their sum is 228. Find the numbers.

26. How many numbers are there between 100 and 1000 that are divisible by 3 or by 11?

27. If it costs $8.40 to cut a piece of pipe into 4 pieces, how much will it cost to cut it into 7 pieces?

28. Find the digit in the 100th place if 5/7 is expressed in decimal form.

29. Joyce wants to make a "necklace" of beads for a decorative bottle that she has. The bottle has a diameter of 10 cm. If each bead is 5 mm long, how many beads will she need?

30. A museum has decided to carpet around three sides of an exhibit (the fourth side will be against the wall. The room is 11 feet by 7 feet and the exhibit is in a cabinet $3\frac{1}{2}$ feet by 2 feet by $7\frac{1}{2}$ feet high). How much carpeting do they need?

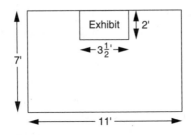

31. If the cabinet housing the exhibit in Problem 30 is all glass, what is the surface area of the cabinet (not including the part against the wall or the part on the floor, since they will not be glass).

32. If the museum is going to put up a rope to keep people 2 feet back from the glass cabinet (in Problems 30 and 31), how much rope will they need?

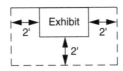

33. What is the measure of a base angle of an isosceles right triangle?

34. If the measures of the angles of a triangle are in the ratio of $1:7:10$, what kind of triangle is it?

35. Can an isosceles triangle have a base angle which is an obtuse triangle? Why or why not?

36. I'm sure you've all heard the saying that "a straight line is the shortest distance between two points." Since a direct path would have to be shorter than going to any other point on the way to the second point, any one side of a triangle would have to be shorter than the sum of the lengths of the other two sides. Using that fact, could a triangle have sides of 4 inches, 11 inches, and 14 inches?

37. Could a triangle have sides of 3 inches, 8 inches, and 4 inches?

38. Find the area of a triangle whose sides have lengths of 3 inches, 4 inches, and 5 inches. (Hint: FIRST, draw the

triangle to scale (and see what you have) and/or study the lengths of the sides and figure out what kind of triangle you have.)

39. Find the area of a triangle whose sides are 4 inches, 5 inches, and 9 inches. (BE CAREFUL WITH THIS ONE!)

40. There is a doorway that is 7 feet by 3 feet. Movers have to bring in a circular mirror which has a diameter of 7½ feet. They find that by tilting it so that it forms the "diagonal" of the doorway, they will have the greatest amount of room. Will they have enough room to bring in the mirror? Explain.

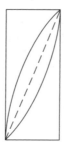

41. 35% of the students in the eighth grade tried out for the school play. There are 320 students in the eighth grade. How many students tried out? If there are 110 parts available, would everyone be able to get some part? Explain.

42. The State Education Department just announced that 87½%, or 2401, students passed the most recent state assessment. How many students took the test?

43. In order to be allowed to remain in business, a cab company must have at least 85% of its cabs less than 2 years old. In the XYZ Cab Company, 225 of its 250 cabs are less than 2 years old. Do they meet the requirement? Will the company be allowed to remain in business?

44. Bill has found that if he plants 72 tomato plants, he will get 162 tomatoes. At that rate, how many tomatoes will he get if he plants 128 plants?

45. The Holiday Card Company has a new promotion. For every 9 cards you buy, you will get 2 cards free. Sandi figured out that over the next few months, she will need

44 cards. If she participates in this promotion, how many will she actually have to buy?

46. A scientist has come up with a theory that it is dangerous to listen to VERY loud music for more than an average of 29 hours per week over any 4-month period. Ron decided to keep track of the number of hours he listened to the radio. He got the following results:

28	38	22	21
18	34	32	25
27	23	39	37
29	29	29	25

Is Ron "safe"?

47. Joe is worried about passing this marking period. So far, he has a passing average, but he is extremely worried about the next test. For the 6 tests he has taken this marking period, he has an average of 68. What is the LOWEST score he could get on the next test, but STILL HAVE A PASSING AVERAGE?

48. I am thinking of 2 numbers. The larger is 10 more than the smaller. Their sum is 182. What are my numbers?

49. The square of an integer is represented by a 3-digit numeral, none of the digits being zero. If the order of the digits is reversed, a **different** numeral is formed, and it also represents the square of an integer. What are the four numerals for which this is true?

50. Find the product in the following multiplication involving (base 10) numerals in which $A \neq B \neq C \neq D \neq E$, and $B = 5$.

$$
\begin{array}{r}
AB \\
\times\ AB \\
\hline
CAB \\
BDB \\
\hline
BEDB
\end{array}
$$

51. What are the last three digits when you find the product?:

$1 \times 2 \times 3 \times 4 \times 5 \times 6 \times 7 \times 8 \times 9 \times 10 \times 11 \times 12 \times 13 \times 14 \times 15$

52. Find the surface area of a cube each of whose edges is 6 inches.

53. An automobile club will "come to the rescue" of any of its members who have a problem within 150 miles of its office. How many square miles do they have to cover?

54. How many feet of fencing will Wendy need for her garden, which is rectangular, and which has dimensions of 27 feet by 11.5 feet?

55. If bulbs are to be placed in the ground so that they have 1 square foot of room, how many bulbs will Wendy need for the garden mentioned in Problem 54?

56. Find the area of the trapezoid pictured below.

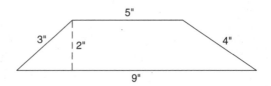

57. Find the perimeter of the trapezoid pictured above.

58. If the measures of the angles of a triangle have measures represented by X, X, and $x + 20$, find the measure of each angle. What kind of triangle is it?

59. Find the area of a right triangle if the hypotenuse is 10 and one leg is 8.

60. The Happy Soap Company took a survey last year to see if people liked their products. Of the 480 people surveyed, 420 spoke very highly of the products. What percent of the people like Happy Soap products?

61. The *Main Street Gazette* decided it would be most profitable if they were to devote 60% of the space in the newspaper to advertising. If there are 80 pages in the newspaper, how many pages will be devoted to advertising?

62. 74% of the students in the sixth grade belong to the Student Organization. If there are 350 students in the sixth grade, how many belong to the S.O.?

63. My favorite football player gained 150 yards on 24 carries last week. If he can keep up that pace, how many yards would he be able to gain this week if he carries the ball 32 times?

64. Find the <u>mean,</u> <u>median,</u> <u>mode,</u> and <u>range</u> of the following set of scores, and then answer the questions that follow.

98	93	100	77	95
92	97	98	97	98
99	80	98	98	90
92	93	96	91	90
98	91	95	90	93

Randi scored a 95 on this test. Is she in the top half of her class?

If the two lowest scores were dropped, by how much would the average increase?

65. Tom is going to be a lifeguard this summer. The pool is shaped like the figure below.

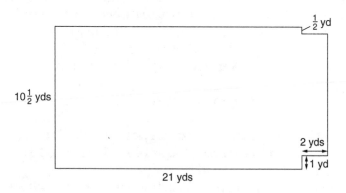

$\frac{1}{2}$ yd

$10\frac{1}{2}$ yds

2 yds

1 yd

21 yds

How many square feet of water does he have to patrol?

66. If there are 185 calories in 2/3 of a cup of James' favorite cereal and he eats 1 cup of it every day for 30 days, how many calories would he have consumed in the month?

67. If an item costs $560, and you have to pay 8% tax, what is the total cost of the item?

68. 85% of the freshman class at State University has arrived for orientation. If 510 students have already arrived, how many are still to come?

69. Your favorite team has won 108 of its 162 games. What is the team's percentage?

70. The length and width of a rectangle are doubled. What percent of the old perimeter is the new? What percent of the old area is the new?

71. How many zeros are there at the (right) end of the product indicated below?

$$1 \times 2 \times 3 \times 4 \times 5 \times 6 \times 7 \times 8 \times 9 \times 10 \times$$
$$11 \times 12 \times 13 \times 14 \times 15 \times 16 \times 17 \times 18 \times 19 \times 20 \times$$
$$21 \times 22 \times 23 \times 24 \times 25 \times 26 \times 27 \times 28 \times 29 \times 30$$

72. (Let's throw in a little Social Studies!) The original colonies represent what percent of the United States as we know it today?

Answers

1. Since the perimeter is 180 feet and each board is 6 **inches,** or **1/2 foot,** wide, **Julian will need 360 boards.**

2. Since the area of the fence is 540 square feet, and one gallon of paint covers 24 square feet, Julian will **need 540/24,** or **22.5 gallons.** (But he may have to BUY 23 gallons.)

3. The area of the living room is 360 square feet. The area of the dining room is 144 square feet. The area of the hallway is 108 square feet. **To carpet this entire area, Ms. Fletcher will need** $(360 + 144 + 108)$ or **612 square feet.**

4. There are 9 square feet in each square yard, so either divide by 9 or change the dimensions to yards. Either way, you find that she needs 68 square yards of carpeting. **The cost is $22.50 × 68, or $1530.**

5. The cost of one 14-inch pizza is the same as the cost of two 7-inch pizzas, so the question really is which gives you more pizza. Since the area of one 14-inch pizza is 154 square inches, and the area of two 7-inch pizzas is 77 square inches, **the one 14-inch pizza is the better buy.** You are getting twice as much pizza for the same price!

6. Find the circumference. **Mr. Dowling would need 62.8 feet of fencing.** If the fencing costs $2.50 a foot,

the fence will cost him $157. The area of his garden will be 314 square feet. If a bag of seed will cover 150 square feet divide 314 by 150, and he will need 2 and a fraction bags of seed. **He will therefore have to BUY 3 bags of seed.** At $11 per bag, **the seed will cost him $33.**

7. The area of the larger, or outer, circle is 314 square inches; the area of the smaller, or inner, circle is 154 square inches. (Use $\pi = 3.14$ in the first case, and 22/7 in the second.) Therefore, the area of the frame is $314 - 154$, or 160 square inches. At $0.75 per square inch, **the cost of refinishing the frame will be $120.**

8.
$$57 \times 2 = 114$$
$$180 - 114 = \mathbf{66°}$$

9.
$$180 - 74 = 106$$
$$106/2 = \mathbf{53°}$$

10. The measures of the angles of the triangle are 15°, 75°, and 90°. Therefore, **it is a right triangle.**

11. Use the Pythagorean Theorem, or recognize that this is a multiple of a 7-24-25 right triangle, and you will find that **the height of the tree is 96 feet.**

12. The test is whether or not $a^2 + b^2 = c^2$. Since they are not equal, **this is not a right triangle.**

13. 90% of the 250 students would be 225. **Since there are only 191 students participating so far, there are not enough students; they will need 34 more.**

14. If 15% of the price is $5250, **the total price of the car is $35,000.** Since she has already paid $5250, **a balance of $29,750 remains.**

15. She **kept** $1500 of the $4000, and that represents 37½%.

16. Solving the proportion, **she will need 1 cup of milk.** (You could also figure that $2/3 = 4/6$, and if **4**/6 serves 4, she will need **6**/6 to serve 6. $6/6 = \mathbf{1}$ cup.

17. **123.75 miles.**

18. Using the price of the 4½-gallon can, we can use a proportion to see that at that rate, 10 gallons should cost

approximately $52.22. But the larger can sells for $52.50. Therefore, **the 4½-gallon can is the better buy.**

19. Not only is the 4½-gallon can a better buy on its own merits, BUT what makes it *an even better buy* is the fact that since Ed needs 45 gallons, he can get 10 cans, or exactly 45 gallons. If he bought the 10-gallon can, he would have to buy 5 cans...even though he does not need 50 gallons.

20. In order to achieve an average of 94 after 6 tests, Ken would need a total of 564 points. His total, so far, is 465 points. Therefore, **Ken would need a grade of 99 on his next test.**

    In order to achieve an average of 95, Ken would need a total of 570 points. Since he has 465 points so far, **he would need 105 points on the next test, and therefore would not be able to accomplish this.**

21. You could find the averages the usual way, OR, now that you've had plenty of practice, I'll show you a shortcut! Since you know the averages you want (90 or 30), you could simply look at "how far off you are." For example, in the case of Amanda, comparing each of her grades to 90, we see she is $+3$, -2, $+7$, and -10. Combining these, we get a total of -2. Therefore, she is 2 points short of achieving the average. (*You could even use this concept to **find** an average. For example, suppose your grades are 93, 96, 87, 93, and 88. "Assume" an average; let's say 90. How far "off" is each grade? You are $+3$, $+6$, -3, $+3$, and -2. Adding these, you get $+7$. Since there are 5 test scores, divide that $+7$ by 5; you get $+1.4$. Now all you have to do is <u>add</u> (since it was positive) <u>1.4</u> to your "assumption," and your average is <u>91.4</u>.*)

    As it happens, in this problem, **no girl achieves the <u>two</u> requirements she needs.**

22. **Mean: 92.866666....**
    Median: 93
    Mode: 97
    Range: 99 − 86 = 13

23. **The numbers are 49 and 34.**

24. Since the remainder is always 1 less than the divisor, it means that we are "1 short" of a multiple. Therefore, find

the LCM and SUBTRACT 1. **The number we are looking for, therefore, is 60 − 1 or <u>59</u>.**

25. **The numbers are 133 and 95.**

26. There are 300 multiples of 3, and 81 multiples of 11. But, there are also 27 multiples of both 3 and 11 or 33. Therefore, **there are 300 + 81 − 27 or 354 numbers satisfying the conditions.**

27. **$16.80.**

28. Since 5/7 = 0.714285, and we are looking for the digit in the 100th place, we simply need to find the 100th place in the cycle. 100/6 = 16 <u>with a remainder of 4</u>. Therefore, we need the digit in the fourth position, or **2.**

29. The circumference is 31.4 cm. Since the beads are 5 **mm**, this is equivalent to .5 cm. Dividing, we find that Joyce will need 62.8, or **63 beads.**

30. The area of the room is 77 square feet; the area taken up by the exhibit is 7 square feet. Therefore, **they need 70 square feet of carpeting.**

31. We have 2 surfaces 2 × 7½, 1 surface which is 3½ × 7½, and 1 surface which is 3½ × 2. The **total surface area that will be in glass is 30 + 26¼ + 7, or 63¼ square inches.**

32. Since we are extending 2 feet in each direction, starting at the left, we have 2 + 2; 2 + 3½ + 2; 2 + 2. Therefore, **they will need 15½ feet of rope.**

33. **45°.**

34. The measures of the angles are 10°, 70°, and 100°. Therefore, **the triangle is obtuse.**

35. **No, since there would have to be 2 angles with the same measure, and that sum would be more than the 180° that we need for all 3 angles.**

36. **Yes, since 4 + 11 > 14.** (You only have to check that the 2 shorter sides have a sum greater than the longest side; the longest plus anything else would always be greater than a shorter side.)

37. **No, since 3 + 4 ≯ 8.**

38. A triangle with sides 3, 4, and 5 is a right triangle. Therefore, the "base" and "height" can be the 2 legs, and we do not have to try to find the length of an altitude we have to draw within the figure.

$A = 1/2\,bh$
$A = 1/2\,(3)(4)$
A = 6 square inches

39. **The area of this triangle is ZERO.** (If you draw it to scale, or use the test we discussed a little earlier, you will see that the 2 shorter sides (since their sum is NOT greater than the third side, but in fact EQUAL) meet at a point ON the third side. We don't really have a triangle at all [unless you want to consider it a VERY flat one!])

40. Using the Pythagorean Theorem, we find that the length of the diagonal of that doorway, and therefore the length of the longest object that will pass through, is approximately 7.62 feet. Since the diameter of the mirror is 7.5, **the mirror will fit through the doorway.**

41. 35% of 320 is **112.** If there are only 110 parts, **there will not be a part for everyone.**

42. **2744 students.**

43. **90% of the cabs are less than 2 years old,** and since all that is required is that 85% fulfill this requirement, **the company fulfills the requirement and will be allowed to remain in business.**

44. **288 tomatoes.**

45. **36 cards.**

46. Ron's average is 28.5; **Ron is safe.**

47. If Joe's average is 68 after 6 tests, he has a total of (68)(6), or 408 points. In order to have a 65 average after 7 tests, he would need a total of 455 points. All he needs for that total would be a score of **47 points.**

48. **86 and 96.**

49. **144; 441; 169; 961.**

50. **5625.**

51. **000** (4 × 5 gives you a 0; 10 gives you a 0; 14 × 15 gives you a 0)

52. The area of each face is 6 × 6, or 36 square inches. There are 6 faces, so **the total surface area of the cube is 36 × 6 or 216 square inches.**

53. **The area of the "circle" covered by the Auto Club is 70,650 square miles.**

54. 77 feet.

55. The area of the garden is 27 × 11.5, or 310.5 square feet. Therefore, there is "room" for 310.5 bulbs, but **since you cannot plant a part of a bulb, Wendy can plant 310 bulbs.**

56. 14 square inches.

57. 21 inches.

58. The angles have measures of 53⅓, 53⅓, and 73⅓. **This is an acute triangle.**

59. If the hypotenuse is 10, and one leg is 8, this is a multiple of a 3-4-5 triangle, and the other leg must be 6. Since we can find the area of a right triangle by using the 2 legs as our base and height, **the area of the triangle is 24 square units.**

60. 87½%.

61. 48 pages.

62. 259 students belong to the Student Organization.

63. 200 yards.

64. Mean: 93.56
Median: 95
Mode: 98
Range: 100 − 77, or 23

65. 238.5 square yards.

66. 8325 calories.

67. The tax is $44.80, so **the total price is $604.80.** (You don't HAVE to find the tax separately in this problem. Since you are asked only for the TOTAL price; that represents the cost of the item (100%) plus the tax (8%), and you can simply find 108% of 560.)

68. If 510 represents 85% of the freshman class, the entire class is 600. Therefore, **there are 90 students still to come.**

69. 66⅔%.

70. If you can't think in general terms, try actual numerical values. You will find that the **perimeter** is doubled, and therefore is **200% of the original,** and the **area** is quadrupled, and therefore is **400% of the original.**

71. There are SEVEN zeros. (4 × 5 gives you 1 zero;
10 gives you 1 zero;
14 × 15 gives you 1 zero;

20 gives you 1 zero;
24 × 25 gives you **2** zeros.)

72. 13/50 = **26%**

AND, the answers to the problems I left hanging during the book (let's see how many you were able to get):
Chapter 6, Problem 8: You can find "the rest" directly by working with "the rest" in terms of percent, so 27 : y = 45/55 and **y = 33.**
Chapter 10: The test for divisibility by 8 involves the last **three** digits being divisible by 8.
Chapter 11: If you multiply by 54, you will get **666,666,666,** since 54 = 9 × 6, and if you want a product of all 5's, multiply 9 × 5 and it means you will have to multiply 12,345,679 by **45.**
And, last but not least, over the course of The Twelve Days of Christmas, **364 gifts are given.**

The pattern looks like this:

1	2	3	4	5	6	7	8	9	10	11	12
	1	2	3	4	5	6	7	8	9	10	11
		1	2	3	4	5	6	7	8	9	10
			1	2	3	4	5	6	7	8	9
				1	2	3	4	5	6	7	8
					1	2	3	4	5	6	7
						1	2	3	4	5	6
							1	2	3	4	5
								1	2	3	4
									1	2	3
										1	2
											1

You can add horizontally, vertically, or even diagonally (all the 1's, all the 2's, etc.).

Index